D0128416

Pre-Algebra Demystified

Demystified Series

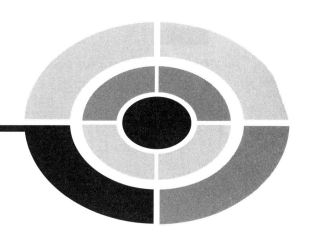

Pre-Algebra Demystified

ALLAN G. BLUMAN

Northern Plains Public Library
Ault Colorado

McGRAW-HILL
New York   Chicago   San Francisco   Lisbon   London
Madrid   Mexico City   Milan   New Delhi   San Juan
Seoul   Singapore   Sydney   Toronto

The McGraw·Hill Companies

¿ing-in-Publication Data is on file with the Library of Congress

Copyright © 2004 by The McGraw-Hill Companies, Inc. All rights reserved. Printed in the United States of America. Except as permitted under the United States Copyright Act of 1976, no part of this publication may be reproduced or distributed in any form or by any means, or stored in a data base or retrieval system, without the prior written permission of the publisher.

5 6 7 8 9 0   DOC/DOC   0 9 8 7 6 5

ISBN 0-07-143931-5

The sponsoring editor for this book was Judy Bass and the production supervisor was Pamela A. Pelton. It was set in Times Roman by Keyword Publishing Services. The art director for the cover was Margaret Webster-Shapiro. Cover design by Handel Low.

Printed and bound by RR Donnelley.

This book was printed on recycled, acid-free paper containing a minimum of 50% recycled, de-inked fiber.

McGraw-Hill books are available at special quantity discounts to use as premiums and sales promotions, or for use in corporate training programs. For more information, please write to the Director of Special Sales, McGraw-Hill Professional, Two Penn Plaza, New York, NY 10121-2298. Or contact your local bookstore.

Information contained in this work has been obtained by The McGraw-Hill Companies, Inc. ("McGraw-Hill") from sources believed to be reliable. However, neither McGraw-Hill nor its authors guarantee the accuracy or completeness of any information published herein, and neither McGraw-Hill nor its authors shall be responsible for any errors, omissions, or damages arising out of use of this information. This work is published with the understanding that McGraw-Hill and its authors are supplying information but are not attempting to render engineering or other professional services. If such services are required, the assistance of an appropriate professional should be sought.

To Brooke Leigh Bluman

CONTENTS

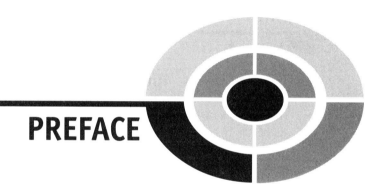

PREFACE

As you know, in order to build a tall building, you need to start with a strong foundation. It is also true in mastering mathematics that you need to start with a strong foundation. This book presents the basic topics in arithmetic and introductory algebra in a logical, easy-to-read format. This book can be used as an independent study course or as a supplement to a pre-algebra course.

To learn mathematics, you must know the vocabulary, understand the rules and procedures, and be able to apply these rules and procedures to mathematical problems in order to solve them. This book is written in a style that will help you with learning. Important terms have been boldfaced, and important rules and procedures have been italicized. Basic facts and helpful suggestions can be found in **Math Notes**. Each section has several worked-out examples showing you how to use the rules and procedures. Each section also contains several practice problems for you to work out to see if you understand the concepts. The correct answers are provided immediately after the problems so that you can see if you have solved them correctly. At the end of each chapter is a 20-question multiple-choice quiz. If you answer most of the problems correctly, you can move on to the next chapter. If not, please repeat the chapter. Make sure that you do not look at the answer before you have attempted to solve the problem.

Even if you know some or all of the material in the chapter, it is best to work through the chapter in order to review the material. The little extra effort will be a great help when you encounter the more difficult material later. After you complete the entire book, you can take the 100-question final exam and determine your level of competence.

Included in this book is a special section entitled "Overcoming Math Anxiety." The author has used it in his workshops for college students who have math anxiety. It consists of three parts. Part 1 explains the nature and some causes of math anxiety. Part 2 explains some techniques that will help you to lessen or eliminate the physical and mental symptoms of math anxiety. Part 3 explains how to succeed in mathematics by using correct study skills. You can read this section before starting the course or any time thereafter.

It is suggested that you do *not* use a calculator since a calculator is only a tool, and there is a tendency to think that if a person can press the right buttons and get the correct answer, then the person *understands* the concepts. This is far from the truth!

Finally, I would like to answer the age-old question, "Why do I have to learn this stuff?" There are several reasons. First, mathematics is used in many academic fields. If you cannot do mathematics, you severely limit your choices of an academic major. Secondly, you may be required to take a standardized test for a job, degree, or graduate school. Most of these tests have a mathematics section. Finally, a working knowledge of arithmetic will go a long way to help you to solve mathematical problems that you encounter in everyday life. I hope this book will help you to learn mathematics.

Best wishes on your success!

ALLAN G. BLUMAN

Acknowledgments

I would like to thank my wife, Betty Claire, for helping me with this project, and I wish to express my gratitude to my editor, Judy Bass, and to Carrie Green for their assistance in the publication of this book.

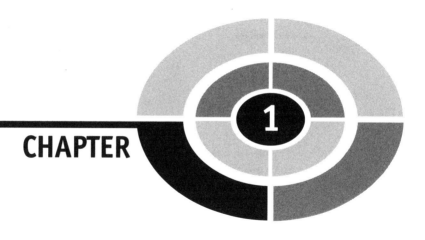

CHAPTER 1

Whole Numbers

Naming Numbers

Our number system is called the **Hindu-Arabic** system or **decimal** system. It consists of 10 symbols or **digits**, 0, 1, 2, 3, 4, 5, 6, 7, 8, and 9, which are used to make our numbers. Each digit in a number has a **place value**. The place value names are shown in Fig. 1-1.

PLACE VALUES														
Trillions			Billions			Millions			Thousands			Ones		
Hundred trillions	Ten trillions	Trillions	Hundred billions	Ten billions	Billions	Hundred millions	Ten millions	Millions	Hundred thousands	Ten thousands	Thousands	Hundreds	Tens	Ones

Fig. 1-1.

In larger numbers, each group of three numbers (called a **period**) is separated by a comma. The names at the top of the columns in Fig. 1-1 are called period names.

To name a number, start at the left and going to the right, read each group of three numbers separately using the period name at the comma. The word "ones" is not used when naming numbers.

EXAMPLE
Name 83,261,403.

SOLUTION
Eighty-three million, two hundred sixty-one thousand, four hundred three.

EXAMPLE
Name 308,657.

SOLUTION
Three hundred eight thousand, six hundred fifty-seven.

EXAMPLE
Name 413,000,000,025.

SOLUTION
Four hundred thirteen billion, twenty-five.

Math Note: Other period names after trillions in increasing order are quadrillion, quintillion, sextillion, septillion, octillion, nonillion, and decillion.

PRACTICE
Name each number:

1. 632
2. 8,506
3. 413,857
4. 33,605,493,600

ANSWERS
1. Six hundred thirty-two
2. Eight thousand, five hundred six
3. Four hundred thirteen thousand, eight hundred fifty-seven

4. Thirty-three billion, six hundred five million, four hundred ninety-three thousand, six hundred

Rounding Numbers

Many times it is not necessary to use an exact number. In this case, an approximate number can be used. Approximations can be obtained by rounding numbers. All numbers can be rounded to specific place values.

To round a number to a specific place value, first locate that place value digit in the number. If the digit to the right of the specific place value digit is 0, 1, 2, 3, or 4, the place value digit remains the same. If the digit to the right of the specific place value digit is 5, 6, 7, 8, or 9, add one to the specific place value digit. All digits to the right of the specific place value digit are changed to zeros.

EXAMPLE
Round 3,261 to the nearest hundred.

SOLUTION
We are rounding to the hundreds place, which is the digit 2. Since the digit to the right of the 2 is 6, add 1 to 2 to get 3. Change all digits to the right to zeros. Hence 3,261 rounded to the nearest hundred is 3,300.

EXAMPLE
Round 38,245 to the nearest thousand.

SOLUTION
We are rounding to the thousands place, which is the digit 8. Since the digit to the right of the 8 is 2, the 8 stays the same. Change all digits to the right of 8 to zeros. Hence 38,245 rounded to the nearest thousand is 38,000.

EXAMPLE
Round 398,261 to the nearest ten thousand.

SOLUTION
We are rounding to the ten thousands place, which is the digit 9. Since the digit to the right of the 9 is 8, the 9 becomes a 10. Then we write the 0 and add the 1 to the next digit. The 3 then becomes a 4. Hence the answer is 400,000.

PRACTICE
1. Round 3,725 to the nearest thousand.

2. Round 563,218 to the nearest ten thousands.
3. Round 80,006 to the nearest 10.
4. Round 478,375 to the nearest hundred thousand.
5. Round 32,864,371 to the nearest million.

ANSWERS
1. 4,000
2. 560,000
3. 80,010
4. 500,000
5. 33,000,000

Addition of Whole Numbers

In mathematics, addition, subtraction, multiplication, and division are called **operations**. The numbers being added are called **addends**. The total is called the **sum.**

$$
\begin{array}{rl}
3 & \leftarrow \text{addend} \\
+\underline{4} & \leftarrow \text{addend} \\
7 & \leftarrow \text{sum}
\end{array}
$$

To add two or more numbers, first write them in a column, and then add the digits in the columns from right to left. If the sum of the digits in any column is 10 or more, write the ones digit and carry the tens digit to the next column and add it to the numbers in that column.

EXAMPLE
Add 24 + 185 + 3,625 + 9.

SOLUTION

$$
\begin{array}{rl}
12 & \leftarrow \text{carry row} \\
24 & \\
185 & \\
3,625 & \\
+\underline{9} & \\
3,843 &
\end{array}
$$

Math Note: To check addition, add from the bottom to the top.

$$
\begin{array}{r}
9 \\
3,625 \\
185 \\
+\ \underline{\ \ 24} \\
3,843
\end{array}
$$

PRACTICE

Add:

1. $43 + 271 + 16$
2. $523 + 8,312 + 53$
3. $6,275,341 + 379,643$
4. $263 + 62,971 + 56 + 483$
5. $3,741 + 60 + 135 + 23,685$

ANSWERS

1. 330
2. 8,888
3. 6,654,984
4. 63,773
5. 27,621

Subtraction of Whole Numbers

In subtraction, the top number is called the **minuend**. The number being subtracted (below the top number) is called the **subtrahend**. The answer in subtraction is called the **remainder** or **difference**.

$$
\begin{array}{rl}
568 & \leftarrow \text{ minuend} \\
-\ \underline{\ 23} & \leftarrow \text{ subtrahend} \\
545 & \leftarrow \text{ difference}
\end{array}
$$

To subtract two numbers, write the numbers in a vertical column and then subtract the bottom digits from the top digits. Proceed from right to left. When the bottom digit is larger than the top digit, borrow from the digit at the top of the next column and add ten to the top digit before subtracting. When borrowing, be sure to reduce the top digit in the next column by 1.

EXAMPLE

Subtract: $16,875 - 3,423$.

SOLUTION

$$
\begin{array}{r}
16{,}875 \\
-\ 3{,}423 \\
\hline
13{,}452
\end{array}
$$

EXAMPLE
Subtract: 3,497 − 659.

SOLUTION

$$
\begin{array}{rcccc}
& 2 & 14 & 8 & 17 \quad \leftarrow \text{ borrowing numbers} \\
& \cancel{3}, & \cancel{4} & \cancel{9} & \cancel{7} \\
- & & 6 & 5 & 9 \\
\hline
& 2, & 8 & 3 & 8
\end{array}
$$

Math Note: To check subtraction, add the difference to the subtrahend to see if you get the minuend.

Check

$$
\begin{array}{r}
3{,}497 \\
-\ \ 659 \\
\hline
2{,}838
\end{array}
\qquad
\begin{array}{r}
2{,}838 \\
+\ \ 659 \\
\hline
3{,}497
\end{array}
$$

PRACTICE
Subtract:

1. 732 − 65
2. 14,375 − 2,611
3. 325,671 − 43,996
4. 12,643,271 − 345,178
5. 7,000,000 − 64,302

ANSWERS
1. 667
2. 11,764
3. 281,675
4. 12,298,093
5. 6,935,698

Multiplication of Whole Numbers

In multiplication, the top number is called the **multiplicand**. The number directly below it is called the **multiplier**. The answer in multiplication is called the **product**. The numbers between the multiplier and the product are called **partial products**.

$$
\begin{array}{r}
3\,4\,2 \\
\times\ \ \underline{3\,4} \\
1\,3\,6\,8 \\
\underline{1\,0\,2\,6} \\
1\,1\,6\,2\,8
\end{array}
\begin{array}{l}
\leftarrow \text{ multiplicand} \\
\leftarrow \text{ multiplier} \\
\leftarrow \text{ partial product} \\
\leftarrow \text{ partial product} \\
\leftarrow \text{ product}
\end{array}
$$

To multiply two numbers when the multiplier is a single digit, write the numbers in a vertical column and then multiply each digit in the multiplicand from right to left by the multiplier. If any of these products is greater than nine, add the tens digit to the product of numbers in the next column.

EXAMPLE
Multiply: 327×6.

SOLUTION

$$
\begin{array}{r}
1\,4 \quad\quad \leftarrow \text{ carry row} \\
\\
3\,2\,7 \\
\times\ \ \underline{6} \\
1,9\,6\,2
\end{array}
$$

To multiply two numbers when the multiplier contains two or more digits, arrange the numbers vertically and multiply each digit in the multiplicand by the right-most digit in the multiplier. Next multiply each digit in the multiplicand by the next digit in the multiplier and place the second partial product under the first partial product, moving one space to the left. Continue the process for each digit in the multiplier and then add the partial products to get the final product.

EXAMPLE
Multiply $2,651 \times 542$.

SOLUTION

$$
\begin{array}{r}
2,6\,5\,1 \\
\times\ \ \ \ 5\,4\,2 \\
\hline
5\,3\,0\,2 \\
1\,0\,6\,0\,4\ \ \ \\
1\,3\,2\,5\,5\ \ \ \ \ \\
\hline
1,4\,3\,6,8\,4\,2
\end{array}
$$

Math Note: To check the multiplication problem, multiply the multiplier by the multiplicand.

$$
\begin{array}{r}
5\,4\,2 \\
\times\ 2\,6\,5\,1 \\
\hline
5\,4\,2 \\
2\,7\,1\,0\ \ \ \\
3\,2\,5\,2\ \ \ \ \\
1\,0\,8\,4\ \ \ \ \ \\
\hline
1,4\,3\,6,8\,4\,2
\end{array}
$$

PRACTICE

Multiply:

1. 28×6
2. 532×49
3. $6,327 \times 146$
4. $52,300 \times 986$
5. $137,264 \times 25$

ANSWERS

1. 168
2. 26,068
3. 923,742
4. 51,567,800
5. 3,431,600

Division of Whole Numbers

In division, the number under the division box is called the **dividend**. The number outside the division box is called the **divisor**. The answer in division is called the **quotient**. Sometimes the answer does not come out *even*; hence, there will be a **remainder**.

$$
\begin{array}{r}
3 \quad \leftarrow \text{quotient} \\
\text{divisor} \ \rightarrow 8\,\overline{)25} \quad \leftarrow \text{dividend} \\
\underline{24} \\
1 \quad \leftarrow \text{remainder}
\end{array}
$$

  The process of long division *consists of a series of steps. They are divide, multiply, subtract, and bring down. When dividing it is also necessary to estimate how many times the divisor divides into the dividend. When the divisor consists of two or more digits, the estimation can be accomplished by dividing the first digit of the divisor into the first one or two digits of the dividend. The process is shown next.*

EXAMPLE
Divide 543 by 37.

SOLUTION
Step 1:

$$
\begin{array}{r}
1 \quad\quad\quad \\
37\,\overline{)543} \quad
\end{array}
$$
        Divide 3 into 5 to get 1

Step 2:

$$
\begin{array}{r}
1 \quad\quad\quad \\
37\,\overline{)543} \quad \\
\underline{37} \quad\quad
\end{array}
$$
        Multiply 1×37

Step 3:

$$
\begin{array}{r}
1 \quad\quad\quad \\
37\,\overline{)543} \quad \\
\underline{37} \quad\quad \\
17 \quad\quad
\end{array}
$$
        Subtract 37 from 54

Step 4:

$$\begin{array}{r} 1 \\ 37\ \overline{)543} \\ 37 \\ \hline 173 \end{array}$$ Bring down 3

Repeat Step 1:

$$\begin{array}{r} 14 \\ 37\ \overline{)543} \\ 37 \\ \hline 173 \end{array}$$ Divide 3 into 17
(Note: The answer 5 is too large, so try 4)

Repeat Step 2:

$$\begin{array}{r} 14 \\ 37\ \overline{)543} \\ 37 \\ \hline 173 \\ 148 \end{array}$$ Multiply 4×37

Repeat Step 3:

$$\begin{array}{r} 14 \\ 37\ \overline{)543} \\ 37 \\ \hline 173 \\ 148 \\ \hline 25 \end{array}$$ Subtract 148 from 173

   Hence, the answer is 14 remainder 25 or 14 R25. Stop when you run out of digits in the dividend to bring down.

EXAMPLE
Divide 2322 by 43.

SOLUTION

$$\begin{array}{r} 54 \\ 43\ \overline{)2322} \\ 215 \\ \hline 172 \\ 172 \\ \hline 0 \end{array}$$

Division can be checked by multiplying the quotient by the divisor and seeing if you get the dividend. For the previous example, multiply 54×43.

$$
\begin{array}{r}
5\,4 \\
\times\ \ 4\,3 \\
\hline
1\,6\,2 \\
2\,1\,6\ \ \\
\hline
2\,3\,2\,2
\end{array}
$$

PRACTICE
Divide:

1. $2666 \div 62$
2. $31{,}568 \div 4$
3. $181{,}044 \div 564$
4. $873 \div 24$
5. $4{,}762 \div 371$

ANSWERS
1. 43
2. 7,892
3. 321
4. 36 R9
5. 12 R310

Word Problems

In order to solve word problems follow these steps:

1. *Read the problem carefully.*
2. *Identify what you are being asked to find.*
3. *Perform the correct operation or operations.*
4. *Check your answer or at least see if it is reasonable.*

In order to know what operation to perform, it is necessary to understand the basic concept of each operation.

ADDITION

When you are asked to find the "sum" or the "total" or "how many in all," and the items are the same in the problem, you add.

EXAMPLE

Find the total calories in a meal consisting of a cheeseburger (550 calories), French fries (352 calories), and a soft drink (110 calories).

SOLUTION

Since you want to find the total number of items and the items are the same (calories), you add: $550 + 352 + 110 = 1012$ calories. Hence, the total number of calories in the meal is 1012.

SUBTRACTION

When you are asked to find the difference, i.e., "how much more," "how much less," "how much larger," "how much smaller," etc., and the items are the same, you subtract.

EXAMPLE

The total area of Lake Superior is 81,000 square miles and the total area of Lake Erie is 32,630 square miles. How much larger is Lake Superior?

SOLUTION

Since you are asked how much larger is Lake Superior than Lake Erie, you subtract. $81,000 - 32,630 = 48,370$. Hence, Lake Superior is 48,370 square miles larger than Lake Erie.

MULTIPLICATION

When you are asked to find a total and the items are different, you multiply.

EXAMPLE

An auditorium consists of 24 rows with 32 seats in each row. How many people can the auditorium seat?

SOLUTION

Since you want a total and the items are different (rows and seats), you multiply:

$$
\begin{array}{r}
3\,2 \\
\times\ \underline{2\,4} \\
1\,2\,8 \\
\underline{6\,4} \\
7\,6\,8
\end{array}
$$

The auditorium will seat 768 people.

DIVISION

When you are given a total and are asked to find how many items in a part, you divide.

EXAMPLE
If 96 calculators are packed in 8 boxes, how many calculators would be placed in each box?

SOLUTION
In this case, the total is 96, and they are to be put into 8 boxes, so to find the answer, divide.

$$
\begin{array}{r}
12 \\
8\,)\overline{96} \\
\underline{8} \\
16 \\
\underline{16} \\
0
\end{array}
$$

Each box would have 12 calculators in it.

PRACTICE
Solve:

1. Find the total number of home runs for the following players: Bonds, 73; McGwire, 70; Sosa, 66; Maris, 61; Ruth, 60.
2. John paid $17,492 for his new automobile. Included in the price was the surround sound package at a cost of $1,293. How much would the automobile have cost without the surround sound package?
3. Dinner for 15 people costs $375. If they decided to split the cost equally, how much would each person pay?
4. A housing developer bought 12 acres of land at $3,517 per acre. What was the total cost of the land?
5. Mark ran a 5-mile race in 40 minutes. About how long did it take him on average to run each mile?

ANSWERS
1. 330 home runs
2. $16,199
3. $25
4. $42,204
5. 8 minutes

Quiz

1. Name 56,254.
   (a) five ten thousands, six thousands, two hundreds, five tens, and four ones
   (b) fifty-six million, two hundred fifty-four
   (c) fifty-six thousand, two hundred fifty-four
   (d) fifty-six hundred, two hundred fifty-four

2. Name 30,000,000,006.
   (a) thirty million, six
   (b) thirty-six billion
   (c) thirty-six million
   (d) thirty billion, six

3. Round 5,164,287 to the nearest hundred thousand.
   (a) 5,200,000
   (b) 5,000,000
   (c) 5,100,000
   (d) 5,160,000

4. Round 879,983 to the nearest hundred.
   (a) 870,000
   (b) 880,983
   (c) 879,000
   (d) 880,000

5. Round 4,321 to the nearest ten.
   (a) 4,300
   (b) 4,320
   (c) 4,000
   (d) 4,330

6. Add $56 + 42 + 165 + 20$.
   (a) 280
   (b) 283
   (c) 293
   (d) 393

7. Add $15,628 + 432,800 + 536$.
   (a) 449,864
   (b) 448,946
   (c) 449,846
   (d) 448,964

8. Add $1,259 + 43,756, + 200 + 8,651$.
    (a) 53,866
    (b) 53,686
    (c) 53,668
    (d) 53,863

9. Subtract $900 - 561$.
    (a) 393
    (b) 933
    (c) 339
    (d) 449

10. Subtract $252,631 - 47,882$.
    (a) 204,749
    (b) 215,859
    (c) 204,947
    (d) 215,895

11. Subtract $371,263,409 - 56,300,266$.
    (a) 341,936,314
    (b) 314,963,143
    (c) 341,832,497
    (d) 314,639,413

12. Multiply 38×52.
    (a) 1,876
    (b) 1,967
    (c) 1,976
    (d) 1,867

13. Multiply 3521×643.
    (a) 2,363,003
    (b) 2,246,003
    (c) 2,642,003
    (d) 2,264,003

14. Multiply 2003×406.
    (a) 800,018
    (b) 803,218
    (c) 813,018
    (d) 813,218

15. Divide $5,481 \div 87$.
    (a) 36
    (b) 73
    (c) 63
    (d) 37 R15

16. Divide $190,314 \div 654$.
    (a) 291
    (b) 290
    (c) 271
    (d) 290 R13

17. Divide $16,177 \div 647$.
    (a) 25
    (b) 27
    (c) 32
    (d) 25 R2

18. In a 30-minute television program, there were 12 1-minute commercials. How many minutes of actual programming were there?
    (a) 42 minutes
    (b) 18 minutes
    (c) 260 minutes
    (d) 28 minutes

19. Tonysha made the following deposits in her savings account: $53, $29, $16, $43, and $35. How much has she saved so far?
    (a) $167
    (b) $166
    (c) $176
    (d) $177

20. South Side College Band purchased 25 new uniforms costing $119 each. What was the total cost?
    (a) $2,975
    (b) $478
    (c) $144
    (d) $3,604

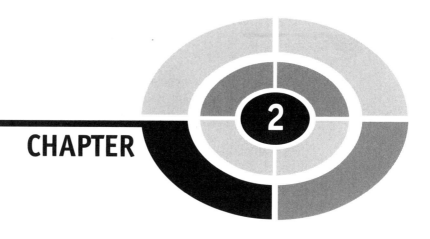

CHAPTER 2

Integers

Basic Concepts

In Chapter 1 we used the set of **whole numbers** which consists of the numbers 0, 1, 2, 3, 4, 5, ... In algebra we extend the set of whole numbers by adding the negative numbers $-1, -2, -3, -4, -5, . . .$. The numbers . . . $-5, -4,$ $-3, -2, -1, 0$ 1, 2, 3, 4, 5, . . . are called **integers**. These numbers can be represented on the number line, as shown in Fig. 2-1. The number zero is called the **origin**.

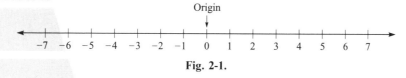

Fig. 2-1.

Math Note: Any number written without a sign (except 0) is considered to be positive; i.e., $6 = +6$. The number zero is neither positive nor negative.

Each integer has an **opposite**. The opposite of a given integer is the corresponding integer, which is exactly the same distance from the origin as the given integer. For example, the opposite of −4 is +4 or 4. The opposite of 0 is 0.

The positive distance any number is from 0 is called the **absolute value** of the number. The symbol for absolute value is | |. Hence, |−6| = 6 and |+10| = 10. In other words, the absolute value of any number except 0 is positive. The absolute value of 0 is 0, i.e., |0| = 0.

Math Note: Do not confuse the concepts of opposite and absolute value. With the exception of zero, to find the opposite of an integer, change its sign and to find the absolute value of an integer, make it positive.

EXAMPLE
Find the opposite of 12.

SOLUTION
The opposite of 12 is −12 since we change the sign.

EXAMPLE
Find |12|.

SOLUTION
|12| = 12 since the absolute value of this number is 12.

EXAMPLE
Find the opposite of −3.

SOLUTION
The opposite of −3 is +3 or 3 since we change the sign.

EXAMPLE
Find |−3|.

SOLUTION
|−3| = 3 since the absolute value is positive.

Sometimes a negative sign is placed outside a number in parentheses. In this case, it means the opposite of the number inside the parentheses. For example, −(−6) means the opposite of −6, which is 6. Hence, −(−6) = 6. Also, −(+8) means the opposite of 8, which is −8. Hence, −(+8) = −8.

EXAMPLE
Find the value of $-(+41)$.

SOLUTION
The opposite of 41 is -41. Hence, $-(41) = -41$.

EXAMPLE
Find the value of $-(-17)$.

SOLUTION
The opposite of -17 is $+17$ or 17. Hence, $-(-17) = 17$.

PRACTICE
1. Find the opposite of -16.
2. Find the opposite of 32.
3. Find $|-23|$.
4. Find $|11|$.
5. Find the opposite of 0.
6. Find $|0|$.
7. Find the value of $-(-10)$.
8. Find the value of $-(+25)$.
9. Find $-(0)$
10. Find the value of $-|-6|$ (Be careful.)

ANSWERS
1. 16
2. -32
3. 23
4. 11
5. 0
6. 0
7. 10
8. -25
9. 0
10. $-|-6| = -(+6) = -6$

Order

When comparing numbers, the symbol $>$ means "greater than." For example, $12 > 3$ is read "twelve is greater than three." The symbol $<$ means "less than." For example, $4 < 10$ is read "four is less than ten."

Given two integers, the number further to the right on the number line is the larger number.

EXAMPLE
Compare −5 with −2.

SOLUTION
Since −2 is further to the right on the number line, it is larger than −5 (see Fig. 2-2). Hence, −5 < −2.

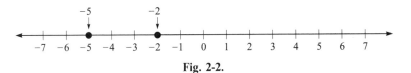

Fig. 2-2.

EXAMPLE
Use > or < to make a true statement.

$$0 \_\_ -3$$

SOLUTION
0 > −3, since 0 is further to the right on the number line.

PRACTICE
Use > or < to make each of the following a true statement:

1. 5 ___ 10
2. −8 ___ −3
3. 0 ___ −6
4. 2 ___ −4
5. −7 ___ −12

ANSWERS
1. <
2. <
3. >
4. >
5. >

Addition of Integers

There are two basic rules for adding integers:

Rule 1: To add two integers with like signs (i.e., both integers are positive or both integers are negative), add the absolute values of the numbers and give the sum the common sign.

EXAMPLE
Add $(+2) + (+4)$.

SOLUTION
Since both integers are positive, add the absolute values of each, $2 + 4 = 6$; then give the answer a $+$ sign. Hence, $(+2) + (+4) = +6$.

EXAMPLE
Add $(-3) + (-2)$.

SOLUTION
Since both integers are negative, add the absolute values $3 + 2 = 5$; then give the answer a $-$ sign. Hence, $(-3) + (-2) = -5$. The rule can be demonstrated by looking at the number lines shown in Fig. 2-3.

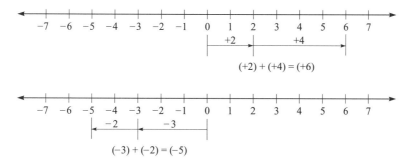

Fig. 2-3.

In the first example, you start at 0 and move two units to the right, ending at $+2$. Then from $+2$, move 4 units to the right, ending at $+6$. Therefore, $(+2) + (+4) = +6$.

In the second example, start at 0 and move 3 units to the left, ending on -3. Then from -3, move 2 units to the left, ending at -5. Therefore, $(-3) + (-2) = -5$.

Rule 2: To add two numbers with unlike signs (i.e., one is positive and one is negative), subtract the absolute values of the numbers and give the answer the sign of the number with the larger absolute value.

EXAMPLE

Add $(+5) + (-2)$.

SOLUTION

Since the numbers have different signs, subtract the absolute values of the numbers, $5 - 2 = 3$. Then give the 3 a positive sign since 5 is larger than 3 and the sign of the 5 is positive. Therefore, $(+5) + (-2) = +3$.

EXAMPLE

Add $(+3) + (-4)$.

SOLUTION

Since the numbers have different signs, subtract the absolute values of the numbers, $4 - 3 = 1$. Then give the 1 a negative sign since 4 is larger than 3. Therefore, $(+3) + (-4) = -1$. This rule can be demonstrated by looking at the number lines shown in Fig. 2-4.

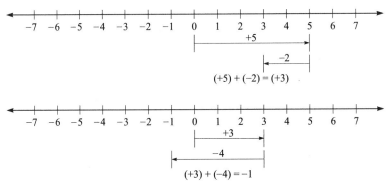

Fig. 2-4.

In the first case, start at 0 and move five units to the right, ending on +5. From there, move 2 units to the left. You will end up at +3. Therefore, $(+5) + (-2) = +3$.

In the second case, start at 0 and move 3 units to the right, ending at +3. From there, move 4 units to the left. You will end on −1. Therefore, $(+3) + (-4) = -1$.

To add three or more integers, you can add two at a time from left to right.

EXAMPLE
Add $(-3) + (-4) + (+2) + (+8) + (-5)$.

SOLUTION

$$
\begin{aligned}
&(-3) + (-4) + (+2) + (+8) + (-5) \\
&= (-7) \quad + (+2) + (+8) + (-5) \\
&= \qquad\qquad (-5) + (+8) + (-5) \\
&= \qquad\qquad\qquad (+3) \quad + (-5) \\
&= \qquad\qquad\qquad\qquad (-2)
\end{aligned}
$$

> **Math Note:** When 0 is added to any number, the answer is the number. For example, $0 + 6 = 6$, $(-3) + 0 = -3$.

PRACTICE
Add:

1. $(+8) + (+2)$
2. $(-5) + (-12)$
3. $(+6) + (-10)$
4. $(-9) + (+4)$
5. $(+6) + (-3)$
6. $(-12) + 0$
7. $(+10) + (-11) + (+1)$
8. $(+3) + (-4) + (+2)$
9. $(-5) + (+12) + (-3) + (+6)$
10. $(-6) + (-8) + (+3) + (+4) + (-2)$

ANSWERS

1. $+10$
2. -17
3. -4
4. -5
5. $+3$
6. -12
7. 0
8. $+1$
9. $+10$
10. -9

Subtraction of Integers

In arithmetic, subtraction is usually thought of as "taking away." For example, if you have six books on your desk and you take four to class, you have two books left on your desk. The "taking away" concepts work well in arithmetic, but with algebra, a new way of thinking about subtraction is necessary.

In algebra, we think of subtraction as adding the opposite. For example, in arithmetic, $8 - 6 = 2$. In algebra, $8 + (-6) = 2$. Notice that in arithmetic, we subtract 6 from 8. In algebra, we add the opposite of 6, which is -6, to 8. In both cases, we get the same answer.

To subtract one number from another, add the opposite of the number that is being subtracted.

EXAMPLE
Subtract $(+12) - (-8)$.

SOLUTION
Add $+8$ (the opposite of -8) to 12 to get 20, as shown.

$$(+12) - (-8) = (+12) + (+8) \leftarrow \text{(opposite of } -8)$$
$$= 20$$

Hence, $(+12) - (-8) = 20$.

EXAMPLE
Subtract $(-6) - (+3)$.

SOLUTION

$$(-6) - (+3) = (-6) + (-3) \leftarrow \text{(opposite of } +3)$$
$$= -9$$

Hence, $(-6) - (+3) = -9$.

Math Note: Sometimes the answers in subtraction do not look correct, but once you get the opposite, you follow the rules of addition. If the two numbers have like signs, use Rule 1 for addition. If the two numbers have unlike signs, use Rule 2 for addition.

EXAMPLE
Subtract $(-10) - (-3)$.

SOLUTION

$$(-10) - (-3) = (-10) + (+3) \leftarrow \text{(opposite of } -3)$$
$$= -7$$

Hence, $(-10) - (-3) = -7$.

EXAMPLE
Subtract $(-8) - (-12)$.

SOLUTION

$$(-8) - (-12) = (-8) + (+12) \text{ (opposite of } -12)$$
$$= +4$$

Hence, $(-8) - (-12) = +4$

PRACTICE
Subtract:

1. $(+10) - (+6)$
2. $(-3) - (-12)$
3. $(-7) - (+18)$
4. $(+15) - (-20)$
5. $(-8) - (-13)$
6. $(+9) - (-2)$
7. $(-11) - (-11)$
8. $(+14) - (-8)$
9. $(-5) - (+5)$
10. $0 - (-3)$

ANSWERS
1. $+4$
2. $+9$
3. -25
4. $+35$
5. $+5$
6. $+11$
7. 0
8. $+22$
9. -10
10. $+3$

Addition and Subtraction

In algebra, the $+$ sign is usually not written when an integer is positive. For example:

$$(+8) + (+2) \quad \text{is written as} \quad 8 + 2$$
$$(-5) - (+6) \quad \text{is written as} \quad -5 - 6$$
$$(+3) - (+8) \quad \text{is written as} \quad 3 - 8$$

When performing the operations of addition and subtraction in the same problem, follow these steps:

- Step 1   Write all the positive signs in front of the positive numbers.
- Step 2   Change all the subtractions to addition (remember to add the opposite).
- Step 3   Add left to right.

EXAMPLE

Perform the indicated operations:

$$3 + (-7) - (-2) + 5 - 12 + 8 - (-6)$$

SOLUTION

Step 1 : $+3 + (-7) - (-2) + (+5) - (+12) + (+8) - (-6)$

Step 2 : $(+3) + (-7) + (+2) + (+5) + (-12) + (+8) + (+6)$

Step 3 : $(+3) + (-7) + (+2) + (+5) + (-12) + (+8) + (+6)$

$$= \quad\quad\quad -4 \quad + (+2) + (+5) + (-12) + (+8) + (+6)$$
$$= \quad\quad\quad\quad\quad -2 \quad\quad + (+5) + (-12) + (+8) + (+6)$$
$$= \quad\quad\quad\quad\quad\quad\quad +3 \quad\quad + (-12) + (+8) + (+6)$$
$$= \quad\quad\quad\quad\quad\quad\quad\quad\quad -9 \quad\quad + (+8) + (+6)$$
$$= \quad\quad\quad\quad\quad\quad\quad\quad\quad\quad\quad\quad -1 \quad\quad + (+6)$$
$$= \quad\quad\quad\quad\quad\quad\quad\quad\quad\quad\quad\quad\quad\quad\quad 5$$

Hence, the answer is 5.

PRACTICE

For each of the following, perform the indicated operations:

1. $-6 + 5 - (-9)$
2. $12 - (-5) - 3$
3. $-18 + 4 - (-7) + 2$
4. $3 + (-4) - (-6)$
5. $-5 + 8 - 6 + 4 - 2 - 3$

ANSWERS
1. 8
2. 14
3. −5
4. 5
5. −4

Multiplication of Integers

For multiplication of integers, there are two basic rules:

Rule 1: To multiply two integers with the same signs, i.e., both are positive or both are negative, multiply the absolute values of the numbers and give the answer a + sign.

EXAMPLE
Multiply $(+8) \times (+2)$.

SOLUTION
Multiply $8 \times 2 = 16$. Since both integers are positive, give the answer a + (positive) sign. Hence, $(+8) \times (+2) = +16$.

EXAMPLE
Multiply $(-9) \times (-3)$.

SOLUTION
Multiply $9 \times 3 = 27$. Since both integers are negative, give the answer a + (positive) sign. Hence, $(-9) \times (-3) = +27$.

Rule 2: To multiply two integers with unlike signs, i.e., one integer is positive and one integer is negative, multiply the absolute values of the numbers and give the answer a − (negative) sign.

EXAMPLE
Multiply $(-7) \times (+6)$.

SOLUTION
Multiply $7 \times 6 = 42$ and give the answer a − (negative) sign. Hence, $(-7) \times (+6) = -42$.

EXAMPLE
Multiply $(+9) \times (-5)$.

SOLUTION
Multiply $9 \times 5 = 45$ and give the answer a $-$ (negative) sign. Hence, $(+9) \times (-5) = -45$.

> **Math Note:** Multiplication can be shown without a times sign. For example, $(-3)(-5)$ means $(-3) \times (-5)$. Also, a dot can be used to represent multiplication. For example, $5 \cdot 3 \cdot 2$ means $5 \times 3 \times 2$.

To multiply three or more non-zero integers, multiply the absolute values and count the number of negative numbers. If there is an odd number of negative numbers, give the answer a $-$ sign. If there is an even number of negative signs, give the answer a $+$ sign.

EXAMPLE
Multiply $(-3)(-4)(-2)$.

SOLUTION
Multiply $3 \times 4 \times 2 = 24$. Since there are 3 negative numbers, the answer is negative. Hence, $(-3)(-4)(-2) = -24$.

EXAMPLE
Multiply $(-5)(+3)(+4)(-2)$.

SOLUTION
Multiply $5 \times 3 \times 4 \times 2 = 120$. Since there are 2 negative numbers, the answer is positive. Hence, $(-5)(+3)(+4)(-2) = +120$.

PRACTICE
Multiply:

1. $(-8)(-5)$
2. $(+6)(-2)$
3. $(+4)(+6)$
4. $(-3)(+5)$
5. $(-4)(+7)$
6. $(-3)(-4)(+2)$
7. $(+3)(-2)(-6)(+9)$
8. $(+10)(+2)(-6)(+9)$
9. $(+8)(-4)(-3)(+8)(-2)(+6)$
10. $(-3)(+10)(-8)(-2)(+4)(-3)$

ANSWERS

1. 40
2. −12
3. 24
4. −15
5. −28
6. 24
7. 324
8. −1080
9. −9216
10. 5760

Division of Integers

Division can be represented in three ways:

1. The division box

$$8\overline{)16}^{2}$$

2. The division sign

$$16 \div 8 = 2$$

3. Fraction

$$\frac{16}{8} = 2$$

The rules for division of integers are the same as the rules for multiplication of integers.

Rule 1: To divide two integers with like signs, divide the absolute values of the numbers and give the answer a + sign.

EXAMPLE
Divide $(+24) \div (+6)$.

SOLUTION
Divide $24 \div 6 = 4$. Since both integers are positive, give the answer a + sign. Hence, $(+24) \div (+6) = +4$.

EXAMPLE
Divide $(-18) \div (-2)$.

SOLUTION
Divide $18 \div 2 = 9$. Since both integers are negative, give the answer a + sign.
Hence, $(-18) \div (-2) = +9$.

> *Rule 2: To divide two integers with unlike signs, divide the absolute values of the integers and give the answer a − sign.*

EXAMPLE
Divide $(-30) \div (+5)$.

SOLUTION
Divide $30 \div 5 = 6$. Since the numbers have unlike signs, give the answer a − sign. Hence, $(-30) \div (+5) = -6$.

EXAMPLE
Divide $(+15) \div (-5)$.

SOLUTION
Divide $15 \div 5 = 3$. Since the numbers have unlike signs, give the answer a − sign. Hence, $(+15) \div (-5) = -3$.

PRACTICE
Divide:

1. $(-63) \div (-9)$
2. $(+45) \div (+5)$
3. $(+38) \div (-2)$
4. $(-20) \div (+10)$
5. $(-30) \div (-5)$

ANSWERS

1. $+7$
2. $+9$
3. -19
4. -2
5. $+6$

Exponents

When the same number is multiplied by itself, the indicated product can be written in **exponential notation**. For example, 3×3 can be written as 3^2; there the 3 is called the **base** and the 2 is called the **exponent**. Also,

$$3 \times 3 \times 3 = 3^3$$

$$3 \times 3 \times 3 \times 3 = 3^4$$

$$3 \times 3 \times 3 \times 3 \times 3 = 3^5, \text{ etc.}$$

3^2 is read as "three squared" or "three to the second power," 3^3 is read as "three cubed" or "three to the third power," 3^4 is read as "three to the fourth power," etc.

Math Note: When no exponent is written with a number, it is assumed to be one. For example, $3 = 3^1$.

EXAMPLE
Find 5^3.

SOLUTION

$$5^3 = 5 \times 5 \times 5 = 125$$

EXAMPLE
Find 2^8.

SOLUTION

$$2^8 = 2 \times 2 \times 2 \times 2 \times 2 \times 2 \times 2 \times 2 = 256$$

Exponents can be used with negative numbers as well. For example, $(-8)^3$ means $(-8) \times (-8) \times (-8)$. Notice that in the case of negative numbers, the integer must be enclosed in parentheses. When the $-$ sign is **not** enclosed in parentheses, it is **not** raised to the power. For example, -8^3 means $-8 \cdot 8 \cdot 8$.

EXAMPLE
Find $(-5)^4$.

SOLUTION
$(-5)^4 = (-5)(-5)(-5)(-5) = 625$

EXAMPLE
Find -5^4.

SOLUTION

$$-5^4 = -5 \times 5 \times 5 \times 5 = -625$$

PRACTICE
Find:

1. 7^4
2. 3^7
3. 5^2
4. 2^1
5. $(-3)^4$
6. $(-6)^3$
7. $(-4)^5$
8. -4^6
9. -7^4
10. -9^3

ANSWERS

1. 2401
2. 2187
3. 25
4. 2
5. 81
6. -216
7. -1024
8. -4096
9. -2401
10. -729

Order of Operations

In the English language, we have punctuation symbols to clarify the meaning of sentences. Consider the following sentence:

John said the teacher is tall.

This sentence could have two different meanings depending on how it is punctuated:

John said, "The teacher is tall."

or

"John," said the teacher, "is tall."

In mathematics, we have what is called an **order of operations** to clarify the meaning when there are operations and grouping symbols (parentheses) in the same problem.

The order of operations is

1. Parentheses
2. Exponents
3. Multiplication or Division, left to right
4. Addition or Subtraction, left to right

Multiplication and division are equal in order and should be performed from left to right. Addition and subtraction are equal in order and should be performed from left to right.

Math Note: The word *simplify* means to perform the operations following the order of operations.

EXAMPLE
Simplify $18 - 6 \times 2 - 16 \div 2$.

SOLUTION

$$
\begin{array}{ll}
18 - 6 \times 2 - 16 \div 2 & \text{Multiplication and division} \\
= \quad 18 - 12 - 8 & \text{Subtraction left to right} \\
= \quad\quad 6 - 8 & \\
= \quad\quad\quad -2 &
\end{array}
$$

EXAMPLE
Simplify $5 + 2^3 - 3 \times 4$.

SOLUTION

$$5 + 2^3 - 3 \times 4 \qquad \text{exponent}$$
$$= 5 + 8 - 3 \times 4 \qquad \text{multiplication}$$
$$= \ 5 + 8 - 12 \qquad \text{addition}$$
$$= \ \ 13 - 12 \qquad \text{subtraction}$$
$$= \ \ \ \ 1$$

EXAMPLE
Simplify $42 - (8 - 6) \times 2^2$.

SOLUTION

$$42 - (8 - 6) \times 2^2 \qquad \text{parentheses}$$
$$= \ \ 42 - 2 \times 2^2 \qquad \text{exponent}$$
$$= \ \ 42 - 2 \times 4 \qquad \text{multiplication}$$
$$= \ \ \ 42 - 8 \qquad \text{subtraction}$$
$$= \ \ \ \ \ 34$$

EXAMPLE
Simplify $19 + (6 - 3)^2 - 18 \div 2$.

SOLUTION

$$19 + (6 - 3)^2 - 18 \div 2 \qquad \text{parentheses}$$
$$= \ 19 + (3)^2 - 18 \div 2 \qquad \text{exponent}$$
$$= \ \ 19 + 9 - 18 \div 2 \qquad \text{division}$$
$$= \ \ \ \ 19 + 9 - 9 \qquad \text{addition}$$
$$= \ \ \ \ \ \ 28 - 9 \qquad \text{subtraction}$$
$$= \ \ \ \ \ \ \ \ 19$$

PRACTICE
Simplify each of the following:

1. $3 - 2^2 + 5$
2. $6 \div 3 \times 2$
3. $32 - (6 \times 2)^2 + 15 \div 3$
4. $8^2 - 3 \times 4 \div 2$
5. $7^3 + (2 \times 5^2 - 4) \div 2$

ANSWERS
1. 4
2. 4

3. −107
4. 58
5. 366

Quiz

1. Find |−12|.
   (a) 12
   (b) −12
   (c) 0
   (d) |12|

2. Find the opposite of 3.
   (a) |3|
   (b) 0
   (c) −3
   (d) 3

3. Find the value of −(−8).
   (a) −8
   (b) 0
   (c) 8
   (d) −(+8)

4. Which number is the largest: −6, −2, 0, 3?
   (a) −6
   (b) −2
   (c) 0
   (d) 3

5. Add −9 + 6.
   (a) −15
   (b) 3
   (c) 15
   (d) −3

6. Add −8 + (−15).
   (a) 23
   (b) −7
   (c) +7
   (d) −23

7. Add $17 + (-3) + (-2) + 5$.
   (a) 17
   (b) 27
   (c) -17
   (d) -20

8. Subtract $-12 - 3$.
   (a) -9
   (b) -15
   (c) 9
   (d) 15

9. Subtract $-5 - (-9)$.
   (a) 4
   (b) 14
   (c) -45
   (d) -14

10. Subtract $18 - (-5)$.
    (a) 13
    (b) 23
    (c) -23
    (d) -13

11. Simplify $8 - 3 + 2 - 7 - 5$.
    (a) 25
    (b) -19
    (c) -5
    (d) $+5$

12. Multiply $(-9)(6)$.
    (a) -54
    (b) -15
    (c) $+54$
    (d) $+15$

13. Multiply $(-7)(+9)$.
    (a) 63
    (b) 72
    (c) -16
    (d) -63

14. Multiply $(-8)(-3)(+4)(-2)(5)$.
    (a) -720
    (b) -960

(c) 540
(d) 400

15. Divide $(-48) \div (+6)$.
   (a) 8
   (b) -8
   (c) 9
   (d) -5

16. Divide $(-22) \div (-11)$.
   (a) -2
   (b) 11
   (c) -11
   (d) 2

17. Find $(-9)^4$.
   (a) -6561
   (b) 36
   (c) 6561
   (d) -36

18. Simplify $8 - 4 \times 2 + 6$.
   (a) 10
   (b) 6
   (c) 12
   (d) 8

19. Simplify $5 \times 2^2 + 4 - 3$.
   (a) 101
   (b) 0
   (c) 19
   (d) 21

20. Simplify $3 \times (6 - 2 + 1) \div 5$.
   (a) 3
   (b) -5
   (c) 2
   (d) 15

21. Simplify $5 \times (13 - 6)^3 + 96 \div 6$.
   (a) 584
   (b) 326
   (c) 1248
   (d) 1731

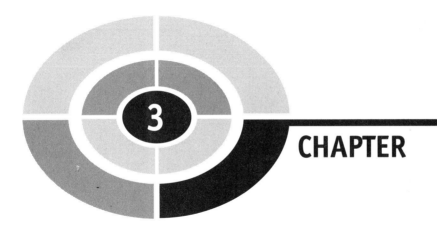

CHAPTER 3

Fractions: Part 1

Basic Concepts

When a whole item is divided into equal parts, the relationship of one or more of the parts to the whole is called a **fraction**. When a pie is cut into six equal parts, each part is one-sixth of the whole pie, as shown in Fig. 3-1.

Fig. 3-1.

The symbol for a fraction is two numbers separated by a bar. The fraction five-eighths is written as $\frac{5}{8}$. The top number of the fraction is called the **numerator** and the bottom number is called the **denominator**.

$$\frac{5}{8} \begin{array}{l} \leftarrow \text{ numerator} \\ \leftarrow \text{ denominator} \end{array}$$

The denominator tells how many parts the whole is being divided into, and the numerator tells how many parts are being used. The fraction $\frac{5}{8}$ means five equal parts of the whole that has been divided into eight equal parts.

> **Math Note:** The denominator of a fraction cannot be zero.

Other fractional parts are shown in Fig. 3-2.

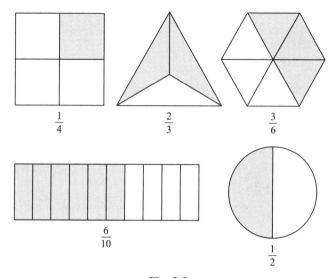

Fig. 3-2.

A fraction whose numerator is less than its denominator is called a **proper** fraction. For example, $\frac{5}{8}$, $\frac{2}{3}$, and $\frac{1}{6}$ are proper fractions. A fraction whose numerator is greater than or equal to its denominator is called an **improper** fraction. For example, $\frac{5}{3}$, $\frac{6}{6}$, and $\frac{10}{4}$ are improper fractions. A number that consists of a whole number and a fraction is called a **mixed** number. For example, $6\frac{5}{9}$, $1\frac{2}{3}$, and $3\frac{1}{8}$ are mixed numbers.

Reducing Fractions

A fraction is said to be in **lowest terms** if both the numerator and denominator cannot be divided evenly by any number except one.

To reduce a fraction to lowest terms, divide the numerator and the denominator by the largest number that divides evenly into both.

EXAMPLE
Reduce $\frac{18}{24}$ to lowest terms.

SOLUTION
Divide both numerator and denominator by 6, as shown:

$$\frac{18}{24} = \frac{18 \div 6}{24 \div 6} = \frac{3}{4}$$

EXAMPLE
Reduce $\frac{15}{25}$ to lowest terms.

SOLUTION
Divide the numerator and denominator by 5, as shown:

$$\frac{15}{25} = \frac{15 \div 5}{25 \div 5} = \frac{3}{5}$$

If the largest number is not obvious, divide the numerator and denominator by any number (except one) that divides into each evenly; then repeat the process until the fraction is in lowest terms.

EXAMPLE
Reduce $\frac{112}{126}$ to lowest terms.

SOLUTION
First divide by 2:

$$\frac{112}{126} = \frac{112 \div 2}{126 \div 2} = \frac{56}{63}$$

Next divide by 7:

$$\frac{56}{63} = \frac{56 \div 7}{63 \div 7} = \frac{8}{9}$$

Math Note: When the numerator of a fraction is zero, the value of the fraction is zero. For example, $\frac{0}{6} = 0$.

PRACTICE
Reduce each of the following fractions to lowest terms:

1. $\dfrac{6}{10}$

2. $\dfrac{4}{12}$

3. $\dfrac{16}{20}$

4. $\dfrac{8}{18}$

5. $\dfrac{150}{200}$

6. $\dfrac{28}{36}$

7. $\dfrac{54}{60}$

8. $\dfrac{160}{216}$

9. $\dfrac{85}{119}$

10. $\dfrac{2000}{2400}$

ANSWERS

1. $\dfrac{3}{5}$

2. $\dfrac{1}{3}$

3. $\dfrac{4}{5}$

4. $\dfrac{4}{9}$

5. $\dfrac{3}{4}$

6. $\dfrac{7}{9}$

7. $\dfrac{9}{10}$

8. $\dfrac{20}{27}$

9. $\dfrac{5}{7}$

10. $\dfrac{5}{6}$

Changing Fractions to Higher Terms

The opposite of reducing fractions is changing fractions to higher terms.

To change a fraction to an equivalent fraction with a larger denominator, divide the larger denominator by the smaller denominator and then multiply the numerator of the fraction with the smaller denominator by the number obtained to get the numerator of the fraction with the larger denominator.

EXAMPLE
Change $\frac{3}{5}$ into an equivalent fraction with a denominator of 30.

SOLUTION
Divide 30 by 5 to get 6; then multiply 3 by 6 to get 18. Hence,

$$\frac{3}{5} = \frac{3 \times 6}{5 \times 6} = \frac{18}{30}$$

EXAMPLE
Change $\frac{5}{8}$ to an equivalent fraction with a denominator of 32.

SOLUTION

$$\frac{5}{8} = \frac{5 \times 4}{8 \times 4} = \frac{20}{32}$$

PRACTICE

1. Change $\dfrac{1}{9}$ to 27ths.

2. Change $\dfrac{3}{8}$ to 48ths.

3. Change $\dfrac{13}{16}$ to 80ths.

4. Change $\dfrac{19}{25}$ to 75ths.

5. Change $\frac{3}{10}$ to 90ths.

ANSWERS

1. $\frac{3}{27}$

2. $\frac{18}{48}$

3. $\frac{65}{80}$

4. $\frac{57}{75}$

5. $\frac{27}{90}$

Changing Improper Fractions to Mixed Numbers

An improper fraction can be changed into an equivalent mixed number. For example, $\frac{13}{4}$ is the same as $3\frac{1}{4}$.

To change an improper fraction to an equivalent mixed number, first divide the numerator by the denominator and then write the answer as a whole number and the remainder as a fraction with the divisor as the denominator and the remainder as the numerator. It may be necessary to reduce the fraction.

EXAMPLE

Change $\frac{17}{5}$ to a mixed number.

SOLUTION

Divide 17 by 5, as shown:

$$
\begin{array}{r}
3 \\
5 \overline{)17} \\
-15 \\
\hline
2
\end{array}
$$

Write the answer as

$$3\tfrac{2}{5}$$

EXAMPLE
Change $\frac{15}{6}$ to a mixed number.

SOLUTION
Divide 15 by 6, as shown:

$$6\overline{)15} \\ \underset{3}{\underline{-12}} 2$$

Write the answer as $2\frac{3}{6}$ and reduce $\frac{3}{6}$ to $\frac{1}{2}$. Hence, the answer is $2\frac{1}{2}$.

Math Note: Any fraction where the numerator is equal to the denominator is always equal to one. For example, $\frac{7}{7} = 1$, $\frac{12}{12} = 1$, etc.

Improper fractions are sometimes equal to whole numbers. This happens when the remainder is zero.

EXAMPLE
Change $\frac{24}{6}$ to a mixed or whole number.

SOLUTION

1. Divide 24 by 6:

$$6\overline{)24} \\ \underset{0}{\underline{-24}} 4$$

2. Write the answer as 4.

PRACTICE
Change each to a mixed number in lowest terms:

1. $\dfrac{6}{5}$

2. $\dfrac{11}{3}$

3. $\dfrac{16}{13}$

4. $\dfrac{70}{15}$

5. $\dfrac{14}{6}$

ANSWERS

1. $1\frac{1}{5}$

2. $3\frac{2}{3}$

3. $1\frac{3}{13}$

4. $4\frac{2}{3}$

5. $2\frac{1}{3}$

Changing Mixed Numbers to Improper Fractions

A mixed number can be changed to an equivalent improper fraction. For example, $5\frac{1}{8}$ is equal to $\frac{41}{8}$.

To change a mixed number to an improper fraction, multiply the whole number by the denominator of the fraction and add the numerator of the fraction to the product. Use this number for the numerator and use the same denominator as the denominator of the improper fraction.

EXAMPLE
Change $7\frac{2}{3}$ to an improper fraction.

SOLUTION
Multiply 7 by 3, then add 2 to get 23, which is the numerator of the improper fraction. Use 3 as the denominator of the improper fraction.

$$7\frac{2}{3} = \frac{3 \times 7 + 2}{3} = \frac{23}{3}$$

EXAMPLE
Change $9\frac{4}{5}$ to an improper fraction.

SOLUTION

$$9\frac{4}{5} = \frac{5 \times 9 + 4}{5} = \frac{49}{5}$$

Math Note: Any whole number can be written as a fraction by placing the whole number over one. For example, $4 = \frac{4}{1}$, $18 = \frac{18}{1}$, etc. Likewise, any fraction with a denominator of one can be written as a whole number: $\frac{5}{1} = 5$, $\frac{7}{1} = 7$, etc.

PRACTICE

Change each of the following mixed numbers to improper fractions:

1. $8\frac{3}{4}$

2. $5\frac{1}{6}$

3. $2\frac{7}{8}$

4. $6\frac{5}{9}$

5. $4\frac{1}{2}$

ANSWERS

1. $\dfrac{35}{4}$

2. $\dfrac{31}{6}$

3. $\dfrac{23}{8}$

4. $\dfrac{59}{9}$

5. $\dfrac{9}{2}$

Quiz

1. What fraction is shown by the shaded portion of Fig. 3-3?

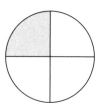

Fig. 3-3.

(a) $\dfrac{1}{2}$

(b) $\dfrac{1}{3}$

(c) $\dfrac{1}{4}$

(d) $\dfrac{3}{4}$

2. What fraction is shown by the shaded portion of Fig. 3-4?

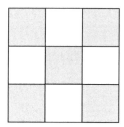

Fig. 3-4.

(a) $\dfrac{4}{5}$

(b) $\dfrac{5}{9}$

(c) $\dfrac{5}{4}$

(d) $\dfrac{4}{9}$

3. Which of the following is a proper fraction?

(a) $\dfrac{6}{5}$

(b) $\dfrac{3}{4}$

(c) $\dfrac{7}{1}$

(d) $\dfrac{3}{3}$

4. Which of the following is an improper fraction?

(a) $\dfrac{5}{8}$

(b) $\dfrac{3}{5}$

(c) $\dfrac{2}{3}$

(d) $\dfrac{9}{7}$

5. Which of the following is a mixed number?

(a) $\dfrac{3}{4}$

(b) $1\frac{5}{6}$

(c) $\dfrac{8}{3}$

(d) $\dfrac{6}{1}$

6. Reduce $\frac{56}{72}$ to lowest terms.

(a) $\dfrac{9}{7}$

(b) $\dfrac{8}{9}$

(c) $\dfrac{7}{8}$

(d) $\dfrac{7}{9}$

7. Reduce $\frac{21}{84}$ to lowest terms.

(a) $\dfrac{1}{4}$

(b) $\dfrac{7}{12}$

(c) $\dfrac{3}{28}$

(d) $\dfrac{4}{1}$

8. Reduce $\frac{120}{200}$ to lowest terms.

(a) $\dfrac{12}{20}$

(b) $\dfrac{3}{5}$

(c) $\dfrac{2}{10}$

(d) $\dfrac{4}{5}$

9. Change $\frac{5}{6}$ to an equivalent fraction in higher terms.

(a) $\dfrac{8}{18}$

(b) $\dfrac{5}{18}$

(c) $\dfrac{10}{18}$

(d) $\dfrac{15}{18}$

10. Change $\frac{3}{4}$ to an equivalent fraction in higher terms.

    (a) $\dfrac{6}{24}$

    (b) $\dfrac{18}{24}$

    (c) $\dfrac{9}{24}$

    (d) $\dfrac{10}{24}$

11. Change $\frac{1}{6}$ to an equivalent fraction in higher terms.

    (a) $\dfrac{7}{42}$

    (b) $\dfrac{8}{42}$

    (c) $\dfrac{13}{42}$

    (d) $\dfrac{21}{42}$

12. Change $\frac{9}{4}$ to a mixed number fraction.

    (a) $2\frac{1}{9}$

    (b) $9\frac{1}{4}$

    (c) $2\frac{1}{4}$

    (d) $5\frac{2}{3}$

13. Change $\frac{7}{4}$ to a mixed number.

    (a) $3\frac{1}{4}$

    (b) $1\frac{3}{4}$

    (c) $4\frac{1}{3}$

(d) $1\frac{7}{11}$

14. Change $5\frac{3}{8}$ to an improper fraction.

(a) $\dfrac{43}{8}$

(b) $\dfrac{23}{8}$

(c) $\dfrac{16}{8}$

(d) $\dfrac{29}{8}$

15. Change $2\frac{7}{8}$ to an improper fraction.

(a) $\dfrac{22}{8}$

(b) $\dfrac{17}{8}$

(c) $\dfrac{23}{8}$

(d) $\dfrac{8}{23}$

16. Change $5\frac{3}{4}$ to an improper fraction.

(a) $\dfrac{4}{23}$

(b) $\dfrac{17}{4}$

(c) $\dfrac{19}{3}$

(d) $\dfrac{23}{4}$

17. Write 8 as a fraction.

(a) $\dfrac{8}{1}$

(b) $\frac{1}{8}$

(c) $\frac{8}{8}$

(d) $\frac{16}{8}$

18. Write $\frac{10}{10}$ as a whole number.
    (a) 10
    (b) 100
    (c) 0
    (d) 1

19. What number is $\frac{0}{7}$ equal to?
    (a) 0
    (b) 7
    (c) 70
    (d) Cannot have 0 in the numerator of a fraction.

20. Which number cannot be used as the denominator of a fraction?
    (a) 1
    (b) 0
    (c) 5
    (d) 100

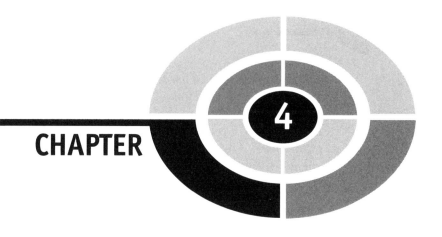

CHAPTER 4

Fractions: Part 2

Finding Common Denominators

In order to add or subtract two or more fractions, they must have the same denominator. This denominator is called a **common denominator**. For any two or more fractions, there are many common denominators; however, in mathematics, we usually use what is called the **lowest (or least) common denominator**, abbreviated **LCD**.

Suppose you wanted to add $\frac{1}{2}$ and $\frac{2}{5}$. Since halves and fifths are different sizes, they cannot be added directly. It is necessary to convert each to equivalent fractions with the same denominator. This can be accomplished by changing each to tenths. Ten is the lowest common denominator of $\frac{1}{2}$ and $\frac{2}{5}$. Since $\frac{1}{2} = \frac{5}{10}$ and $\frac{2}{5} = \frac{4}{10}$, the two fractions can now be added as $\frac{5}{10} + \frac{4}{10} = \frac{9}{10}$.

There are several methods of finding the lowest common denominator. The easiest method is to simply look at the numbers in the denominator of the fractions and "see" what is the smallest number that all the denominator numbers divide into evenly. For example, 2 and 5 both divide into 10 evenly. However, this only works when the denominators are small numbers.

Another method is to list the multiples of the number in the denominator and eventually you will find a common multiple which is the same number as a common denominator.

A **multiple** of a given number is the product of the given number and any other whole number. Multiples of a given number are obtained by multiplying the given number by 0, 1, 2, 3, 4, 5, etc. For example, the multiples of 5 are

$$5 \times 0 = 0$$
$$5 \times 1 = 5$$
$$5 \times 2 = 10$$
$$5 \times 3 = 15$$
$$5 \times 4 = 20$$
$$5 \times 5 = 25$$
etc.

The multiples of 6 are

$$6 \times 0 = 0$$
$$6 \times 1 = 6$$
$$6 \times 2 = 12$$
$$6 \times 3 = 18$$
$$6 \times 4 = 24$$
$$6 \times 5 = 30$$
etc.

Now, if you want to find a common denominator, simply list the multiples of the numbers in the denominators until a common multiple of both numbers is found. For example, the common denominator of $\frac{1}{5}$ and $\frac{1}{6}$ is found as follows:

$$0, 5, 10, 15, 20, 25, 30,$$
$$0, 6, 12, 18, 24, 30,$$

Since 30 is the smallest common multiple of 5 and 6, it is the lowest common denominator of $\frac{1}{5}$ and $\frac{1}{6}$.

When the denominators of the fractions are large, another method, called the **division method**, can be used. The procedure for finding the lowest common denominator of two or more fractions is as follows:

- *Step 1   Arrange the numbers in the denominators in a row.*
- *Step 2   Divide by the smallest number that divides evenly into two or more of the numbers.*
- *Step 3   Bring down to the next row all quotients and remaining numbers not used.*
- *Step 4   Continue dividing until no two of the denominators can be divided evenly by any number other than one.*
- *Step 5   Multiply all the divisors and the remaining numbers to get the lowest common denominator.*

EXAMPLE

Find the lowest common denominator of the fractions $\frac{5}{16}$, $\frac{3}{8}$, and $\frac{1}{24}$.

SOLUTION

Arrange the numbers in a row and start dividing, as shown:

$$
\begin{array}{r|rrr}
2) & 16 & 8 & 24 \\
\hline
2) & 8 & 4 & 12 \\
\hline
2) & 4 & 2 & 6 \\
\hline
 & 2 & 1 & 3
\end{array}
$$

After dividing by 2 three times, there are no two numbers in the last row that can be divided by any number other than one, so you stop. Multiply the divisors and the numbers in the bottom row to get the lowest common denominator. $2 \times 2 \times 2 \times 2 \times 1 \times 3 = 48$. Hence, the lowest common denominator of the fractions $\frac{5}{16}$, $\frac{3}{8}$, and $\frac{1}{24}$ is 48.

EXAMPLE

Find the lowest common denominator of $\frac{3}{20}$, $\frac{2}{15}$, and $\frac{7}{30}$.

SOLUTION

$$
\begin{array}{r|rrr}
2) & 20 & 15 & 30 \\
\hline
3) & 10 & 15 & 15 \\
\hline
5) & 10 & 5 & 5 \\
\hline
 & 2 & 1 & 1
\end{array}
$$

Hence, the LCD is $2 \times 3 \times 5 \times 2 \times 1 \times 1 = 60$.

PRACTICE

Find the LCD of each of the following fractions:

1. $\frac{1}{4}$ and $\frac{3}{10}$

2. $\frac{7}{20}$ and $\frac{13}{24}$

3. $\frac{9}{10}$ and $\frac{5}{18}$

4. $\frac{3}{16}$, $\frac{7}{20}$, and $\frac{5}{28}$

5. $\frac{2}{5}$, $\frac{5}{18}$, and $\frac{3}{20}$

ANSWERS

1. 20
2. 120
3. 90
4. 560
5. 180

Addition of Fractions

In order to add two or more fractions, the fractions must have the same denominator.

To add two or more fractions:

- *Step 1   Find the LCD of the fractions.*
- *Step 2   Change the fractions to equivalent fractions in higher terms with the LCD.*
- *Step 3   Add the numerators of the fractions.*
- *Step 4   Reduce or simplify the answer if possible.*

EXAMPLE

Add $\frac{1}{10} + \frac{7}{10}$.

SOLUTION

Since both fractions have the same denominator, add the numerators and reduce the fractions:

$$\frac{1}{10} + \frac{7}{10} = \frac{1+7}{10} = \frac{8}{10} = \frac{4}{5}$$

EXAMPLE
Add $\frac{3}{8} + \frac{5}{6}$.

SOLUTION
Step 1   Find the LCD. It is 24.
Step 2   Change the fractions to higher terms:

$$\frac{3}{8} = \frac{9}{24} \qquad \frac{5}{6} = \frac{20}{24}$$

Step 3   Add the numerators:

$$\frac{9}{24} + \frac{20}{24} = \frac{9+20}{24} = \frac{29}{24}$$

Step 4   Simplify the answers:

$$\frac{29}{24} = 1\frac{5}{24}$$

EXAMPLE
Add $\frac{2}{3} + \frac{5}{6} + \frac{3}{4}$.

SOLUTION
Addition can be done vertically, as shown:

$$\frac{2}{3} = \frac{8}{12}$$

$$\frac{5}{6} = \frac{10}{12}$$

$$+ \frac{3}{4} = \frac{9}{12}$$

$$\frac{27}{12} = 2\frac{3}{12} = 2\frac{1}{4}$$

PRACTICE
Add each of the following fractions:

1. $\frac{5}{11} + \frac{3}{11}$

2. $\dfrac{7}{10}+\dfrac{2}{3}+\dfrac{1}{5}$

3. $\dfrac{1}{4}+\dfrac{5}{6}+\dfrac{1}{3}$

4. $\dfrac{5}{12}+\dfrac{3}{8}+\dfrac{1}{6}$

5. $\dfrac{9}{16}+\dfrac{3}{4}+\dfrac{1}{2}$

ANSWERS

1. $\dfrac{8}{11}$

2. $1\frac{17}{30}$

3. $1\frac{5}{12}$

4. $\dfrac{23}{24}$

5. $1\frac{13}{16}$

Addition of Mixed Numbers

To add mixed numbers, first add the fractions and then add the whole numbers; next simplify the answer, if necessary.

EXAMPLE

Add $5\frac{2}{3}+3\frac{1}{8}+2\frac{5}{6}$.

SOLUTION

$$5\tfrac{2}{3}=5\tfrac{16}{24}$$

$$3\tfrac{1}{8}=3\tfrac{3}{24}$$

$$+\ 2\tfrac{5}{6}=2\tfrac{20}{24}$$

$$10\tfrac{39}{24}=10+1+\tfrac{5}{8}=11\tfrac{5}{8}$$

EXAMPLE
Add $4\frac{2}{3} + 9\frac{1}{10} + 7\frac{3}{5}$.

SOLUTION

$$4\frac{2}{3} = 4\frac{20}{30}$$
$$9\frac{1}{10} = 9\frac{3}{30}$$
$$+\ 7\frac{3}{5} = 7\frac{18}{30}$$
$$20\frac{41}{30} = 20 + 1 + \frac{11}{30} = 21\frac{11}{30}$$

PRACTICE
Add:

1. $5\frac{2}{3} + 3\frac{1}{6}$
2. $8\frac{5}{12} + 18\frac{3}{8}$
3. $31\frac{7}{16} + 14\frac{5}{6}$
4. $5\frac{1}{2} + 8\frac{5}{8} + 6\frac{3}{4}$
5. $21\frac{3}{8} + 9\frac{11}{12} + 31\frac{5}{16}$

ANSWERS

1. $8\frac{5}{6}$
2. $26\frac{19}{24}$
3. $46\frac{13}{48}$
4. $20\frac{7}{8}$
5. $62\frac{29}{48}$

Subtraction of Fractions

The rule for subtracting fractions is the same as the rule for adding fractions.
To subtract two fractions:

- *Step 1   Find the LCD of the fractions.*
- *Step 2   Change the fractions to equivalent fractions in higher terms with the LCD.*
- *Step 3   Subtract the numerators of the fractions.*
- *Step 4   Reduce the answer if possible.*

EXAMPLE
Subtract $\frac{9}{10} - \frac{3}{4}$.

SOLUTION

$$\frac{9}{10} = \frac{18}{20}$$

$$-\ \frac{3}{4} = \frac{15}{20}$$

$$\frac{3}{20}$$

EXAMPLE
Subtract $\frac{11}{12} - \frac{3}{8}$.

SOLUTION

$$\frac{11}{12} = \frac{22}{24}$$

$$-\ \frac{3}{8} = \frac{9}{24}$$

$$\frac{13}{24}$$

PRACTICE
Subtract:

1. $\dfrac{3}{8} - \dfrac{1}{8}$

2. $\dfrac{5}{6} - \dfrac{1}{2}$

3. $\dfrac{14}{15} - \dfrac{3}{20}$

4. $\dfrac{4}{5} - \dfrac{7}{12}$

5. $\dfrac{7}{8} - \dfrac{1}{3}$

ANSWERS

1. $\dfrac{1}{4}$

2. $\dfrac{1}{3}$

3. $\dfrac{47}{60}$

4. $\dfrac{13}{60}$

5. $\dfrac{13}{24}$

Subtraction of Mixed Numbers

Subtraction of mixed numbers is a little more complicated than addition of mixed numbers since it is sometimes necessary to **borrow**.

To subtract two mixed numbers:

- *Step 1   Find the LCD of the fractions.*
- *Step 2   Change the fractions to higher terms with the LCD.*
- *Step 3   Subtract the fractions, borrowing if possible.*
- *Step 4   Subtract the whole numbers.*
- *Step 5   Reduce or simplify the answer if necessary.*

When the fraction in the subtrahend is smaller than the fraction in the minuend, borrowing is not necessary.

EXAMPLE
Subtract $9\frac{3}{4} - 5\frac{1}{8}$.

SOLUTION:

$$9\tfrac{3}{4} = 9\tfrac{6}{8}$$
$$-\ 5\tfrac{1}{8} = 5\tfrac{1}{8}$$
$$\overline{\phantom{-\ 5\tfrac{1}{8} = }\ 4\tfrac{5}{8}}$$

Principles of Borrowing

When the fraction in the subtrahend is larger than the fraction in the minuend, it is necessary to borrow from the whole number.

When borrowing is necessary, take one (1) away from the whole number, change it to a fraction with the same numerator and denominator, and then add it to the fraction.

EXAMPLE
Borrow 1 from $8\frac{1}{5}$.

SOLUTION

$$8\tfrac{1}{5} = 7 + 1 + \frac{1}{5} \qquad \text{borrow 1 from 8}$$

$$= 7 + \frac{5}{5} + \frac{1}{5} \qquad \text{change 1 to } \frac{5}{5}$$

$$= 7\tfrac{6}{5} \qquad\qquad \text{add } \frac{5}{5} + \frac{1}{5}$$

EXAMPLE
Borrow 1 from $6\frac{3}{8}$.

SOLUTION

$$6\tfrac{3}{8} = 5 + 1 + \frac{3}{8}$$

$$= 5 + \frac{8}{8} + \frac{3}{8}$$

$$= 5\tfrac{11}{8}$$

The next examples show how to use borrowing when subtracting mixed numbers.

EXAMPLE
Subtract $8\frac{2}{5} - 4\frac{2}{3}$.

SOLUTION

$$8\tfrac{2}{5} = 8\tfrac{6}{15} = 7\tfrac{21}{15}$$

$$\underline{-4\tfrac{2}{3} = 4\tfrac{10}{15} = 4\tfrac{10}{15}}$$

$$3\tfrac{11}{15}$$

EXAMPLE
Subtract $12\frac{1}{8} - 2\frac{3}{4}$.

SOLUTION

$$12\frac{1}{8} = 12\frac{1}{8} = 11\frac{9}{8}$$
$$\underline{-2\frac{3}{4} = \ 2\frac{6}{8} = \ 2\frac{6}{8}}$$
$$9\frac{3}{8}$$

EXAMPLE
Subtract $9 - 6\frac{1}{5}$.

SOLUTION

$$9 = 8\frac{5}{5}$$
$$\underline{-6\frac{1}{5} = 6\frac{1}{5}}$$
$$2\frac{4}{5}$$

PRACTICE
Subtract:

1. $6\frac{7}{16} - 3\frac{5}{16}$
2. $16\frac{7}{8} - 1\frac{3}{5}$
3. $8\frac{5}{6} - 5\frac{1}{3}$
4. $14\frac{3}{8} - 11\frac{1}{12}$
5. $15\frac{3}{4} - 10$
6. $5\frac{2}{7} - 2\frac{5}{7}$
7. $11\frac{5}{16} - 5\frac{3}{4}$
8. $21\frac{3}{8} - 14\frac{5}{6}$
9. $8\frac{3}{5} - 7\frac{5}{6}$
10. $10 - 1\frac{2}{3}$

ANSWERS

1. $3\frac{1}{8}$

2. $15\frac{11}{40}$

3. $3\frac{1}{2}$

4. $3\frac{7}{24}$

5. $5\frac{3}{4}$

6. $2\frac{4}{7}$

7. $5\frac{9}{16}$

8. $6\frac{13}{24}$

9. $\dfrac{23}{30}$

10. $8\frac{1}{3}$

Multiplication of Fractions

Multiplying fractions uses the principle of **cancellation**. Cancellation saves you from reducing the answer after multiplying.

To cancel, divide any numerator and any denominator by the largest number possible (that is, the greatest common factor).

To multiply two or more fractions:

- *Step 1   Cancel if possible.*
- *Step 2   Multiply numerators.*
- *Step 3   Multiply denominators.*
- *Step 4   Simplify the answer if possible.*

EXAMPLE
Multiply $\frac{3}{8} \times \frac{4}{9}$.

SOLUTION

$$\frac{3}{8} \times \frac{4}{9} = \frac{\overset{1}{\cancel{3}}}{\underset{2}{\cancel{8}}} \times \frac{\overset{1}{\cancel{4}}}{\underset{3}{\cancel{9}}}$$

$$= \frac{1 \times 1}{2 \times 3}$$

$$= \frac{1}{6}$$

EXAMPLE
Multiply $\frac{5}{8} \times \frac{2}{3} \times \frac{4}{5}$

SOLUTION

$$\frac{5}{8} \times \frac{2}{3} \times \frac{4}{5} = \frac{\overset{1}{\cancel{5}}}{\underset{2}{\underset{1}{\cancel{8}}}} \times \frac{\overset{1}{\cancel{2}}}{3} \times \frac{\overset{1}{\cancel{4}}}{\underset{1}{\cancel{5}}}$$

$$= \frac{1 \times 1 \times 1}{1 \times 3 \times 1}$$

$$= \frac{1}{3}$$

PRACTICE
Multiply:

1. $\dfrac{3}{16} \times \dfrac{8}{15}$

2. $\dfrac{9}{10} \times \dfrac{5}{12}$

3. $\dfrac{2}{21} \times \dfrac{7}{8}$

4. $\dfrac{2}{3} \times \dfrac{3}{4} \times \dfrac{5}{6}$

5. $\dfrac{1}{2} \times \dfrac{6}{7} \times \dfrac{2}{3}$

ANSWERS

1. $\dfrac{1}{10}$

2. $\dfrac{3}{8}$

3. $\dfrac{1}{12}$

4. $\dfrac{5}{12}$

5. $\dfrac{2}{7}$

Multiplication of Mixed Numbers

Multiplying mixed numbers is similar to multiplying fractions.

To multiply two or more mixed numbers, first change the mixed numbers to improper fractions and then follow the steps for multiplying fractions.

EXAMPLE
Multiply $1\frac{3}{5} \times 3\frac{3}{4}$.

SOLUTION

$$1\tfrac{3}{5} \times 3\tfrac{3}{4} = \frac{8}{5} \times \frac{15}{4}$$

$$= \frac{\overset{2}{\cancel{8}}}{\underset{1}{\cancel{5}}} \times \frac{\overset{3}{\cancel{15}}}{\underset{1}{\cancel{4}}}$$

$$= \frac{2 \times 3}{1 \times 1}$$

$$= \frac{6}{1} = 6$$

Math Note: When a whole number is used in a multiplication problem, make it into a fraction with a denominator of 1.

EXAMPLE
Multiply $9 \times 2\frac{2}{3}$.

SOLUTION

$$9 \times 2\tfrac{2}{3} = 9 \times \frac{8}{3}$$

$$= \frac{9}{1} \times \frac{8}{3}$$

$$= \frac{\overset{3}{\cancel{9}}}{1} \times \frac{8}{\underset{1}{\cancel{3}}}$$

$$= \frac{3 \times 8}{1 \times 1}$$

$$= \frac{24}{1} = 24$$

EXAMPLE
Multiply $1\frac{3}{10} \times 4\frac{1}{6} \times 3\frac{3}{4}$.

SOLUTION

$$1\tfrac{3}{10} \times 4\tfrac{1}{6} \times 3\tfrac{3}{4} = \frac{13}{10} \times \frac{25}{6} \times \frac{15}{4}$$

$$= \frac{13}{\underset{2}{\cancel{10}}} \times \frac{\overset{5}{\cancel{25}}}{\underset{2}{\cancel{6}}} \times \frac{\overset{5}{\cancel{15}}}{4}$$

$$= \frac{13 \times 5 \times 5}{2 \times 2 \times 4}$$

$$= \frac{325}{16}$$

$$= 20\tfrac{5}{16}$$

PRACTICE
Multiply:

1. $1\frac{1}{8} \times 2\frac{2}{3}$
2. $3\frac{5}{6} \times 2\frac{2}{5}$
3. $15 \times 1\frac{7}{10}$
4. $2\frac{1}{2} \times 1\frac{3}{5} \times 3\frac{1}{4}$
5. $4\frac{1}{8} \times 6\frac{2}{3} \times 1\frac{1}{5}$

ANSWERS

1. 3
2. $9\frac{1}{5}$
3. $25\frac{1}{2}$
4. 13
5. 33

Division of Fractions

When dividing fractions, it is necessary to use the **reciprocal** of a fraction. To find the reciprocal of a fraction, interchange the numerator and the denominator. For example, the reciprocal of $\frac{2}{3}$ is $\frac{3}{2}$. The reciprocal of $\frac{5}{8}$ is $\frac{8}{5}$. The reciprocal of 15 is $\frac{1}{15}$. Finding the reciprocal of a fraction is also called **inverting**.

To divide two fractions, invert the fraction after the division sign and multiply.

EXAMPLE
Divide $\frac{3}{4} \div \frac{7}{8}$.

SOLUTION

$$\frac{3}{4} \div \frac{7}{8} = \frac{3}{4} \times \frac{8}{7} \quad \text{invert } \frac{7}{8} \text{ and multiply}$$

$$= \frac{3}{\cancel{4}_{1}} \times \frac{\cancel{8}^{2}}{7}$$

$$= \frac{6}{7}$$

EXAMPLE
Divide $\frac{5}{6} \div \frac{2}{3}$.

SOLUTION

$$\frac{5}{6} \div \frac{2}{3} = \frac{5}{6} \times \frac{3}{2}$$

$$= \frac{5}{\cancel{6}} \times \frac{\cancel{3}^{1}}{2}$$
$$_{2}$$

$$= \frac{5}{4} = 1\frac{1}{4}$$

PRACTICE
Divide:

1. $\dfrac{5}{8} \div \dfrac{1}{4}$

2. $\dfrac{7}{10} \div \dfrac{21}{30}$

3. $\dfrac{2}{9} \div \dfrac{1}{3}$

4. $\dfrac{3}{11} \div \dfrac{9}{22}$

5. $\dfrac{1}{8} \div \dfrac{7}{12}$

ANSWERS

1. $2\frac{1}{2}$

2. 1

3. $\dfrac{2}{3}$

4. $\dfrac{2}{3}$

5. $\dfrac{3}{14}$

Division of Mixed Numbers

To divide mixed numbers, change them into improper fractions, invert the fraction after the division sign, and multiply.

EXAMPLE
Divide $3\frac{3}{4} \div 1\frac{7}{8}$.

SOLUTION

$$3\frac{3}{4} \div 1\frac{7}{8} = \frac{15}{4} \div \frac{15}{8}$$

$$= \frac{15}{4} \times \frac{8}{15}$$

$$= \frac{\overset{1}{\cancel{15}}}{\underset{1}{\cancel{4}}} \times \frac{\overset{2}{\cancel{8}}}{\underset{1}{\cancel{15}}}$$

$$= \frac{2}{1} = 2$$

EXAMPLE
Divide $2\frac{5}{6} \div 4\frac{1}{3}$.

SOLUTION

$$2\frac{5}{6} \div 4\frac{1}{3} = \frac{17}{6} \div \frac{13}{3}$$

$$= \frac{17}{6} \times \frac{3}{13}$$

$$= \frac{17}{\underset{2}{\cancel{6}}} \times \frac{\overset{1}{\cancel{3}}}{13}$$

$$= \frac{17}{26}$$

PRACTICE
Divide:

1. $7\frac{1}{2} \div 3\frac{3}{4}$
2. $1\frac{1}{8} \div 5\frac{1}{12}$
3. $6\frac{3}{5} \div 2\frac{7}{10}$
4. $9\frac{1}{3} \div 4\frac{5}{8}$
5. $4\frac{2}{5} \div 1\frac{9}{10}$

ANSWERS

1. 2
2. $\dfrac{27}{122}$
3. $2\frac{4}{9}$
4. $2\frac{2}{111}$
5. $2\frac{6}{19}$

Word Problems

Word problems involving fractions are performed using the same procedure as those involving whole numbers.

EXAMPLE
A plumber needs to install a sink using pieces of pipe measuring $6\frac{3}{4}$ inches, $10\frac{2}{3}$ inches, $5\frac{1}{6}$ inches, and $3\frac{1}{8}$ inches. Find the total length of the pipe needed to cut all the pieces.

SOLUTIONS
Since we are finding a total, we use addition:

$$
\begin{array}{rcl}
6\frac{3}{4} & = & 6\frac{18}{24} \\
10\frac{2}{3} & = & 10\frac{16}{24} \\
5\frac{1}{6} & = & 5\frac{4}{24} \\
+3\frac{1}{8} & = & 3\frac{3}{24} \\
\hline
& & 24\frac{41}{24} = 25\frac{17}{24} \text{ inches}
\end{array}
$$

The plumber would need a piece of pipe at least $25\frac{17}{24}$ inches long.

EXAMPLE
It takes Bill $2\frac{3}{4}$ hours to cut a lawn. It takes Walter $3\frac{1}{8}$ hours to cut the same lawn. How much faster is Bill?

SOLUTION
Since we are looking for a difference, we subtract:

$$
\begin{aligned}
3\frac{1}{8} = 3\frac{1}{8} = \ \ &2\frac{9}{8} \\
- \ 2\frac{3}{4} = 2\frac{6}{8} = -&2\frac{6}{8} \\
\hline
&\frac{3}{8}
\end{aligned}
$$

It takes Bill $\frac{3}{8}$ of an hour less to cut the lawn.

EXAMPLE
How much will a trip of 212 miles cost if the allowance is $27\frac{3}{4}$ cents per mile?

SOLUTION
Multiply:

$$212 \times 27\frac{3}{4} = \frac{212}{1} \times \frac{111}{4}$$

$$= \frac{\overset{53}{\cancel{212}}}{1} \times \frac{111}{\cancel{4}}$$

$$= 53 \times 111$$

$$= 5883 \text{ cents}$$

To change cents to dollars, divide by 100.

$$5883 \div 100 = \$58.83$$

Hence the allowance for the trip is $58.83.

EXAMPLE
How many cubic feet will a 240-gallon tank hold if one cubic foot of water is about $7\frac{1}{2}$ gallons?

SOLUTION
Divide:

$$240 \div 7\tfrac{1}{2} = 240 \div \frac{15}{2}$$

$$= \frac{\overset{16}{\cancel{240}}}{1} \times \frac{2}{\underset{1}{\cancel{15}}}$$

$$= 16 \times 2$$
$$= 32 \text{ cubic feet}$$

Hence, the tank will hold about 32 cubic feet of water.

PRACTICE
1. A recipe calls for $2\tfrac{1}{4}$ cups of flour. If the baker wishes to make the cake 4 times as large as the original recipe, how many cups of flour should she use?
2. A $3\tfrac{1}{2}$ inch bolt is placed through a piece of wood that is $1\tfrac{7}{8}$ inches thick. How much of the bolt is extending out?
3. In order to make a support beam, a carpenter nails together a piece of wood that is $1\tfrac{1}{2}$ inches thick with a piece of wood that is $2\tfrac{1}{4}$ inches thick. How thick is the support?
4. A sign is to be $2\tfrac{3}{8}$ inches wide. How many signs can be cut from a piece of card stock that is $11\tfrac{7}{8}$ inches wide?
5. A person decided to buy an automobile that costs \$19,500 with $\tfrac{2}{5}$ paid in cash and the balance paid in 12 monthly installments. How much is the down payment and each monthly payment?

ANSWERS
1. 9 cups
2. $1\tfrac{5}{8}$ inches
3. $3\tfrac{3}{4}$ inches
4. 5
6. \$7800; \$975

Comparing Fractions

To compare two or more fractions:

- *Step 1   Find the LCD of the fractions.*
- *Step 2   Change the fractions to higher terms using the LCD.*
- *Step 3   Compare the numerators.*

EXAMPLE
Which is larger: $\frac{3}{8}$ or $\frac{4}{7}$?

SOLUTION

$$\frac{3}{8} = \frac{21}{56} \quad \text{and} \quad \frac{4}{7} = \frac{32}{56}$$

Since 32 is larger than 21, the fraction $\frac{4}{7}$ is larger than $\frac{3}{8}$.

PRACTICE
Which is larger:

1. $\frac{11}{16}$ or $\frac{5}{8}$?

2. $\frac{2}{5}$ or $\frac{3}{7}$?

3. $\frac{7}{10}$ or $\frac{3}{4}$

4. $\frac{8}{9}$ or $\frac{5}{6}$?

5. $\frac{3}{8}$ or $\frac{2}{3}$?

ANSWERS

1. $\frac{11}{16}$

2. $\frac{3}{7}$

3. $\frac{3}{4}$

4. $\dfrac{8}{9}$

5. $\dfrac{2}{3}$

Operations with Positive and Negative Fractions

Every fraction has three signs: the sign of the number in the numerator, the sign of the number in the denominator, and the sign in front of the fraction (called the sign of the fraction). For example,

$$\frac{3}{4} = +\frac{+3}{+4}$$

When a fraction is negative, such as $-\frac{3}{4}$, the negative sign can be placed in the numerator or denominator. For example,

$$-\frac{3}{4} = \frac{-3}{4} = \frac{3}{-4}$$

When performing operations with fractions that are negative, keep the negative sign with the numerator of the fraction and perform the operations.

EXAMPLE
Add $-\frac{2}{3} + \frac{3}{4}$.

SOLUTION

$$-\frac{2}{3} + \frac{3}{4} = \frac{-2}{3} + \frac{3}{4}$$

$$= \frac{-8}{12} + \frac{9}{12}$$

$$= \frac{-8+9}{12}$$

$$= \frac{1}{12}$$

EXAMPLE
Subtract $-\frac{11}{12} - \frac{3}{8}$.

SOLUTION

$$-\frac{11}{12} - \frac{3}{8} = \frac{-11}{12} - \frac{3}{8}$$

$$= \frac{-22}{24} - \frac{9}{24}$$

$$= \frac{-22 - 9}{24}$$

$$= \frac{-31}{24} = -\frac{31}{24} = -1\frac{7}{24}$$

EXAMPLE
Multiply $\frac{3}{8} \times \left(-\frac{4}{9}\right)$.

SOLUTION

$$\frac{3}{8} \times \left(-\frac{4}{9}\right) = \frac{3}{8} \times \frac{-4}{9}$$

$$= \frac{\overset{1}{\cancel{3}}}{\underset{2}{\cancel{8}}} \times \frac{\overset{1}{\cancel{-4}}}{\underset{3}{\cancel{9}}}$$

$$= \frac{-1}{6} = -\frac{1}{6}$$

EXAMPLE
Divide $\frac{2}{3} \div \left(-\frac{5}{8}\right)$.

SOLUTION

$$\frac{2}{3} \div \left(-\frac{5}{8}\right) = \frac{2}{3} \times \frac{-8}{5}$$

$$= \frac{2 \times (-8)}{3 \times 5}$$

$$= \frac{-16}{15} = -1\frac{1}{15}$$

PRACTICE

Perform the following indicated operations:

1. $-\dfrac{1}{2}+\left(-\dfrac{3}{4}\right)+\dfrac{2}{3}$

2. $\dfrac{9}{10}-\left(-\dfrac{3}{5}\right)$

3. $\dfrac{1}{8}\times\left(-\dfrac{5}{6}\right)\times\left(-\dfrac{4}{5}\right)$

4. $\dfrac{13}{24}\div\left(-\dfrac{2}{3}\right)$

5. $-\dfrac{3}{4}\times\dfrac{2}{5}-\dfrac{1}{8}$

ANSWERS

1. $-\dfrac{7}{12}$

2. $1\frac{1}{2}$

3. $\dfrac{1}{12}$

4. $-\dfrac{13}{16}$

5. $-\dfrac{17}{40}$

Quiz

1. Add $\dfrac{11}{12}+\dfrac{5}{8}$.

   (a) $1\frac{5}{6}$

   (b) $1\frac{3}{4}$

(c) $1\frac{13}{24}$

(d) $2\frac{1}{8}$

2. Subtract $\frac{9}{10} - \frac{5}{6}$.

(a) $\dfrac{1}{15}$

(b) $\dfrac{7}{60}$

(c) 1

(d) $\dfrac{11}{30}$

3. Multiply $\frac{5}{6} \times \frac{13}{15}$.

(a) $\dfrac{18}{23}$

(b) $\dfrac{13}{18}$

(c) $\dfrac{17}{24}$

(d) $\dfrac{67}{90}$

4. Divide $\frac{1}{6} \div \frac{2}{9}$.

(a) $\dfrac{2}{53}$

(b) $1\frac{1}{9}$

(c) $\dfrac{1}{27}$

(d) $\dfrac{3}{4}$

5. Add $6\frac{3}{4} + 3\frac{5}{8} + 2\frac{1}{3}$.

(a) $11\frac{5}{6}$

(b) $12\frac{17}{24}$

(c) $12\frac{5}{12}$

(d) $11\frac{7}{24}$

6. Subtract $15\frac{2}{9} - 6\frac{5}{12}$.

   (a) $8\frac{29}{36}$

   (b) $9\frac{7}{12}$

   (c) $8\frac{7}{36}$

   (d) $9\frac{11}{12}$

7. Multiply $3\frac{1}{8} \times 2\frac{5}{6}$.

   (a) $6\frac{5}{48}$

   (b) $5\frac{23}{24}$

   (c) $7\frac{11}{24}$

   (d) $8\frac{41}{48}$

8. Divide $9\frac{1}{5} \div 6\frac{3}{4}$.

   (a) $2\frac{4}{15}$

   (b) $3\frac{3}{20}$

   (c) $1\frac{17}{60}$

   (d) $1\frac{49}{135}$

9. Add $5\frac{1}{2} + \frac{3}{4} + 1\frac{7}{8}$.

   (a) $6\frac{3}{4}$

   (b) $8\frac{1}{8}$

   (c) $7\frac{5}{8}$

   (d) $9\frac{5}{16}$

10. $\frac{5}{12} \times \frac{2}{3} + 1\frac{7}{8}$.

    (a) $3\frac{7}{24}$

    (b) $4\frac{3}{8}$

    (c) $2\frac{11}{72}$

    (d) $1\frac{5}{6}$

11. Find the cost of $2\frac{1}{4}$ lbs of bananas at 60¢ per pound.
    (a) 125¢
    (b) 145¢
    (c) 155¢
    (d) 135¢

12. How many $\frac{1}{2}$-pound bags can be filled with salt from a container which holds $26\frac{1}{2}$ pounds?
    (a) 53
    (b) $13\frac{1}{4}$
    (c) 27
    (d) 52

13. A person bought $8\frac{2}{3}$ ounces of caramel candy, $5\frac{1}{8}$ ounces of jelly candy and $9\frac{1}{4}$ ounces of licorice candy. How many ounces did the person buy altogether?

    (a) $22\frac{5}{24}$ ounces

    (b) $23\frac{7}{24}$ ounces

    (c) $22\frac{13}{24}$ ounces

    (d) $23\frac{1}{24}$ ounces

14. Last month Madison weighed $134\frac{3}{4}$ pounds. This month she weighed $129\frac{2}{3}$ pounds. How much weight did she lose?
    (a) $6\frac{2}{3}$ pounds
    (b) $5\frac{1}{12}$ pounds
    (c) $5\frac{1}{2}$ pounds
    (d) $4\frac{1}{8}$ pounds

15. Which fraction is the smallest: $\dfrac{5}{16}, \dfrac{3}{8}, \dfrac{1}{4}, \dfrac{1}{3}$?

   (a) $\dfrac{5}{16}$

   (b) $\dfrac{3}{8}$

   (c) $\dfrac{1}{4}$

   (d) $\dfrac{1}{3}$

16. Add $-\dfrac{7}{8} + \left(-\dfrac{3}{5}\right) + \dfrac{9}{10}$.

   (a) $2\frac{3}{8}$

   (b) $1\frac{7}{40}$

   (c) $-\dfrac{23}{40}$

   (d) $-\dfrac{1}{3}$

17. Subtract $-1\frac{1}{8} - 5\frac{2}{3}$.

   (a) $4\frac{13}{24}$

   (b) $6\frac{19}{24}$

   (c) $-6\frac{19}{24}$

   (d) $-4\frac{13}{24}$

18. Multiply $-\dfrac{3}{4} \times \left(-\dfrac{1}{6}\right) \times \left(\dfrac{2}{3}\right)$.

   (a) $\dfrac{1}{12}$

   (b) $\dfrac{6}{13}$

   (c) $-\dfrac{6}{13}$

   (d) $-\dfrac{1}{12}$

19. Divide $-6\frac{7}{8} \div \left(-2\frac{1}{2}\right)$.

    (a) $-3\frac{1}{6}$

    (b) $12\frac{1}{8}$

    (c) $17\frac{3}{16}$

    (d) $2\frac{3}{4}$

20. Perform the indicated operations: $6\frac{1}{4} + \left(-\frac{2}{3}\right) \times 1\frac{1}{2}$

    (a) $5\frac{1}{4}$

    (b) $10\frac{3}{8}$

    (c) $7\frac{1}{4}$

    (d) $4\frac{1}{12}$

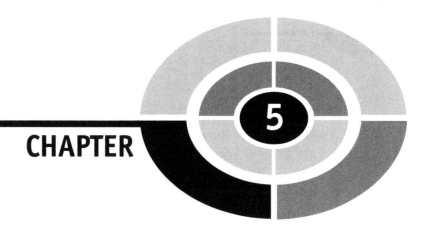

CHAPTER 5

Decimals

Naming Decimals

As with whole numbers, each digit of a decimal has a **place value**. The place value names are shown in Fig. 5-1.

PLACE VALUES					
Tenths	Hundredths	Thousandths	Ten thousandths	Hundred thousandths	Millionths

Fig. 5-1.

For example, in the number 0.6235, the 2 is in the hundredths place. The 5 is in the ten thousandths place.

When naming a decimal, read the number from left to right as you would read a whole number, and then use the place value name for the last digit of the number.

EXAMPLE
Name 0.6235.

SOLUTION
First write in words 6235, and then write the place value of the five after it:

Six thousand two hundred thirty-five ten thousandths.

EXAMPLE
Name 0.000006.

SOLUTION

Six millionths.

When there is a whole number and a decimal, the decimal part is written using the word "and."

EXAMPLE
Name 57.263.

SOLUTION

Fifty-seven and two hundred sixty-three thousandths.

PRACTICE
Name each number:

1. 0.0032
2. 0.444
3. 0.87236
4. 39.61
5. 182.9

ANSWERS
1. Thirty-two ten thousandths
2. Four hundred forty-four thousandths
3. Eighty-seven thousand two hundred thirty-six hundred thousandths
4. Thirty-nine and sixty-one hundredths
5. One hundred eighty-two and nine tenths

Rounding Decimals

Decimals are rounded the same way as whole numbers are rounded.

To round a decimal to a specific place value, first locate that place value digit in the number. If the digit to the right is 0, 1, 2, 3, or 4, the place value digit remains the same. If the digit to the right of the specific value digit is 5, 6, 7, 8, or 9, add one to the specific place value digit. All digits to the right of the place value digit are dropped.

EXAMPLE
Round 0.53871 to the nearest thousandth.

SOLUTION
We are rounding to the thousandths place, which is the digit 8. Since the digit to the right of the 8 is 7, raise the 8 to a 9 and drop all digits to the right of the 9. Hence, the answer is 0.539.

EXAMPLE
Round 83.2146 to the nearest hundredth.

SOLUTION
The digit in the hundredths place is 1 and since the digit to the right of 1 is 4, the 1 remains the same. Hence, the answer is 83.21.

PRACTICE
1. Round 0.38 to the nearest tenth
2. Round 0.2324 to the nearest hundredth
3. Round 26.4567 to the nearest thousandth
4. Round 0.8999 to the nearest ten thousandths
5. Round 2.812 to the nearest one

ANSWERS
1. 0.4
2. 0.23
3. 26.457
4. 1
5. 3

> **Math Note:** Zeros can be affixed to the end of a decimal on the right side of the decimal point. For example, 0.62 can be written as 0.620 or 0.6200. Likewise, the zeros can be dropped if they are at the end of a decimal on the right side of the decimal point. For example, 0.3750 can be written as 0.375.

Addition of Decimals

In order to add two or more decimals, write the numbers in a column placing the decimal points of each number in a vertical line. Add the numbers and place the decimal point in the sum directly under the other decimal points above.

EXAMPLE
Add 3.2 + 56.87 + 381.561.

SOLUTION

$$
\begin{array}{r}
3.200 \\
56.870 \\
+381.561 \\
\hline
441.631
\end{array}
$$

Zeros are annexed to keep the columns straight

EXAMPLE
Add 47.315 + 126.2 + 0.03.

SOLUTION

$$
\begin{array}{r}
47.315 \\
126.200 \\
+\ \ \ 0.030 \\
\hline
173.545
\end{array}
$$

PRACTICE
Add:

1. 0.63 + 2.5 + 8.332
2. 43.7 + 126.21 + 0.561
3. 29.1 + 32 + 627.52 + 41.6
4. 18.003 + 0.0006 + 43.1
5. 92.7 + 3.562 + 4362.81

ANSWERS
1. 11.462
2. 170.471
3. 730.22
4. 61.1036
5. 4459.072

Subtraction of Decimals

Subtracting decimals is similar to adding decimals.

   To subtract two decimals, write the decimals in a column, placing the decimal points in a vertical line. Subtract the numbers and place the decimal point in the difference directly under the other decimal points.

EXAMPLE
Subtract 43.871 − 15.23.

SOLUTION

$$\begin{array}{r} 43.871 \\ -15.230 \\ \hline 28.641 \end{array}$$ Annex a zero to keep the columns straight

EXAMPLE
Subtract 18.3 − 9.275.

SOLUTION

$$\begin{array}{r} 18.300 \\ -\ 9.275 \\ \hline 9.025 \end{array}$$ Annex zeros to keep the columns straight

PRACTICE
Subtract:

1. 19.356 − 14.124
2. 237.6 − 153.48
3. 0.833 − 0.6241
4. 7.8 − 3.45
5. 2.63175 − 0.4111

ANSWERS
1. 5.232
2. 84.12
3. 0.2089
4. 4.35
5. 2.22065

Multiplication of Decimals

To multiply two decimals, multiply the two numbers, disregarding the decimal points, and then count the total number of digits to the right of the decimal points in the two numbers. Count the same number of places from the right in the product and place the decimal point there. If there are fewer digits in the product than places, prefix as many zeros as needed.

EXAMPLE
Multiply 32.1 × 0.62.

SOLUTION

```
      32.1        Three decimal places are needed in the answer
   × 0.62
      642
    1926
   19.902
```

EXAMPLE
Multiply 51.3 × 0.0006.

SOLUTION

```
        51.3
   × 0.0006
    0.03078
```

Since five places are needed in the answer, it is necessary to use one zero in front of the product.

PRACTICE
Multiply:

1. 17.82×6.3
2. 321.2×2.03
3. 0.003×0.06
4. 5.02×0.4
5. 723.4×0.615

ANSWERS
1. 112.266
2. 652.036
3. 0.00018
4. 2.008
5. 444.891

Division of Decimals

When dividing two decimals, it is important to find the correct location of the decimal point in the quotient. There are two cases.

Case 1. To divide a decimal by a whole number, divide as though both numbers were whole numbers and place the decimal point in the quotient directly above the decimal point in the dividend.

EXAMPLE
Divide 179.2 by 56.

SOLUTION

$$
\begin{array}{r}
3.2 \\
56\overline{)179.2} \\
\underline{168} \\
112 \\
\underline{112} \\
0
\end{array}
$$

EXAMPLE
Divide 38.27 by 43.

SOLUTION

$$
\begin{array}{r}
.89 \\
43\overline{)38.27} \\
\underline{344} \\
387 \\
\underline{387} \\
0
\end{array}
$$

Sometimes it is necessary to place zeros in the quotient.

EXAMPLE
Divide 0.0035 by 7.

SOLUTION

$$
\begin{array}{r}
.0005 \\
7\overline{)0.0035} \\
\underline{35} \\
0
\end{array}
$$

Case 2. When the divisor contains a decimal point, move the point to the right of the last digit in the divisor. Then move the point to the right the same number of places in the dividend. Divide and place the point in the quotient directly above the point in the dividend.

EXAMPLE
Divide 14.973 by 0.23.

SOLUTION

$$0.23\overline{)14.973}$$

Move the point two places to the right as shown:

$$
\begin{array}{r}
65.1 \\
23\overline{)1497.3} \\
\underline{138} \\
117 \\
\underline{115} \\
23 \\
\underline{23} \\
0
\end{array}
$$

Sometimes it is necessary to place zeros in the dividend.

EXAMPLE
Divide 18 by 0.375.

SOLUTION

$$0.375 \overline{)18}$$

Move the point three places to the right after annexing three zeros:

$$
\begin{array}{r}
48 \\
375 \overline{)18000.} \\
\underline{1500} \\
3000 \\
\underline{3000} \\
0
\end{array}
$$

PRACTICE
1. $113.4 \div 6$
2. $24.486 \div 33$
3. $1.15938 \div 0.54$
4. $9.5 \div 4.75$
5. $0.0224 \div 28$

ANSWERS
1. 18.9
2. 0.742
3. 2.147
4. 2
5. 0.0008

Comparing Decimals

To compare two or more decimals, place the numbers in a vertical column with the decimal points in a straight line with each other. Add zeros to the ends of the decimals so that they all have the same number of decimal places. Then compare the numbers, ignoring the decimal points.

EXAMPLE
Which is larger, 0.43 or 0.561?

SOLUTION

$$
\begin{array}{ccccc}
0.43 & & 0.430 & & 430 \\
& \rightarrow & & \rightarrow & \\
0.561 & & 0.561 & & 561
\end{array}
$$

Since 561 is larger than 430, 0.561 is larger than 0.43.

EXAMPLE

Arrange the decimals 0.63, 0.256, 3.1, and 0.8 in order of size, smallest to largest.

SOLUTION

0.630	630
0.256	256
3.100	3100
0.800	800

In order: 0.256, 0.63, 0.8, and 3.1.

PRACTICE
1. Which is larger: 0.85 or 0.623?
2. Which is smaller: 0.003 or 0.02?
3. Arrange in order (smallest first): 0.44, 0.4, 0.444.
4. Arrange in order (largest first): 0.565, 0.51, 0.5236.
5. Arrange in order (largest first): 0.01, 0.1, 0.001, 1.0.

ANSWERS
1. 0.85
2. 0.003
3. 0.4, 0.44, 0.444
4. 0.565, 0.5236, 0.51
5. 1.0, 0.1, 0.01, 0.001

Word Problems

To solve a word problem involving decimals, follow the procedure given in Chapter 1:

1. Read the problem carefully.
2. Identify what you are being asked to find.
3. Perform the correct operation or operations.
4. Check your answer or at least see if it is reasonable.

EXAMPLE

Normal body temperature is 98.6°F. If a man has a fever and his temperature is 101.2°F, how much higher is his temperature than normal?

SOLUTION

$$
\begin{array}{ll}
\begin{array}{r} 101.2 \\ -\ 98.6 \end{array} &
\begin{array}{ll} \text{Check:} & \begin{array}{r} 98.6 \\ +\ 2.6 \\ \hline 101.2 \end{array} \end{array}
\end{array}
$$

Hence, his temperature is 2.6°F higher than normal.

EXAMPLE

If an automobile averages 26.3 miles per gallon, how many gallons of gasoline will be needed for a trip that is a distance of 341.9 miles?

SOLUTION

Divide 341.9 by 26.3:

$$
26.3\overline{)341.9} \qquad
\begin{array}{r}
13 \\
263\overline{)3419} \\
\underline{263} \\
789 \\
\underline{789} \\
0
\end{array}
\qquad
\begin{array}{ll}
\text{Check:} & \begin{array}{r} 26.3 \\ \times\ \ 13 \\ \hline 789 \\ \underline{263} \\ 341.9 \end{array}
\end{array}
$$

Hence, 13 gallons will be needed.

PRACTICE

1. During the month of November, it snowed on six occasions. The amounts were 2.1 in., 1.87 in., 0.3 in., 0.95 in., 1.2 in., and 1.95 in. Find the total amount of snow during November.
2. Find the cost of making 154 copies at $0.07 per page.
3. A person borrowed $2688 to pay for a purchase. If it was to be paid back in 12 equal monthly payments over 4 years, find the monthly payment. No interest was charged.
4. A nurse withdrew 10.25 ml of medication from a bottle containing 16.3 ml. How much medication was left in the bottle?
5. In a state lottery, 8 people divided a prize of $20,510.40. How much was each person's share?

ANSWERS

1. 8.37 in.
2. $10.78

3. $56
4. 6.05 ml
5. $2,563.80

Changing Fractions to Decimals

A fraction can be converted to an equivalent decimal. For example $\frac{1}{4} = 0.25$. When a fraction is converted to a decimal, it will be in one of two forms: a **terminating decimal** or a **repeating decimal**.

To change a fraction to a decimal, divide the numerator by the denominator.

EXAMPLE
Change $\frac{3}{4}$ to a decimal.

SOLUTION

$$
\begin{array}{r}
.75 \\
4\overline{)3.00} \\
\underline{28} \\
20 \\
\underline{20} \\
0
\end{array}
$$

Hence, $\frac{3}{4} = 0.75$.

EXAMPLE
Change $\frac{5}{8}$ to a decimal.

SOLUTION

$$
\begin{array}{r}
.625 \\
8\overline{)5.000} \\
\underline{48} \\
20 \\
\underline{16} \\
40 \\
\underline{40} \\
0
\end{array}
$$

Hence, $\frac{5}{8} = 0.625$.

EXAMPLE

Change $\frac{3}{11}$ to a decimal.

SOLUTION

$$
\begin{array}{r}
.2727 \\
11\overline{)3.0000} \\
\underline{22} \\
80 \\
\underline{77} \\
30 \\
\underline{22} \\
80
\end{array}
$$

Hence, $\frac{3}{11} = 0.2727\ldots$

The repeating decimal can be written as $0.\overline{27}$.

EXAMPLE

Change $\frac{5}{6}$ to a decimal.

SOLUTION

$$
\begin{array}{r}
.833 \\
6\overline{)5.000} \\
\underline{48} \\
20 \\
\underline{18} \\
20 \\
\underline{18} \\
2
\end{array}
$$

Hence, $\frac{5}{6} = 0.833\ldots$ or $0.8\overline{3}$.

A mixed number can be changed to a decimal by first changing it to an improper fraction and then dividing the numerator by the denominator.

EXAMPLE

Change $6\frac{2}{5}$ to a decimal.

SOLUTION

$6\frac{2}{5} = \frac{32}{5}$

$$
\begin{array}{r}
6.4 \\
5\overline{)32.0} \\
\underline{30} \\
20 \\
\underline{20} \\
0
\end{array}
$$

Hence, $6\frac{2}{5} = 6.4$.

PRACTICE

Change each of the following fractions to a decimal:

1. $\dfrac{3}{8}$

2. $\dfrac{1}{6}$

3. $\dfrac{11}{20}$

4. $\dfrac{5}{12}$

5. $4\frac{1}{3}$

ANSWERS

1. 0.375
2. $0.1\overline{6}$
3. 0.55
4. $0.41\overline{6}$
5. $4.\overline{3}$

Changing Decimals to Fractions

To change a terminating decimal to a fraction, drop the decimal point and place the digits to the right of the decimal in the numerator of a fraction whose denominator corresponds to the place value of the last digit in the decimal. Reduce the answer if possible.

EXAMPLE

Change 0.8 to a fraction.

SOLUTION

$$0.8 = \frac{8}{10} = \frac{4}{5}$$

Hence, $0.8 = \dfrac{4}{5}$.

EXAMPLE

Change 0.45 to a fraction.

SOLUTION

$$0.45 = \frac{45}{100} = \frac{9}{20}$$

Hence, $0.45 = \frac{9}{20}$.

EXAMPLE
Change 0.0035 to a fraction.

SOLUTION

$$0.0035 = \frac{35}{10,000} = \frac{7}{2,000}$$

Hence, $0.0035 = \frac{7}{2,000}$.

PRACTICE
Change each of the following decimals to a reduced fraction:

1. 0.7
2. 0.06
3. 0.45
4. 0.875
5. 0.0015

ANSWERS

1. $\frac{7}{10}$

2. $\frac{3}{50}$

3. $\frac{9}{20}$

4. $\frac{7}{8}$

5. $\frac{3}{2000}$

Changing a repeating decimal to a fraction requires a more complex procedure, and this procedure is beyond the scope of this book. However, Table 5-1 can be used for some common repeating decimals.

$\frac{1}{12} = 0.08\overline{3}$	$\frac{1}{6} = 0.1\overline{6}$	$\frac{1}{3} = 0.\overline{3}$	$\frac{5}{12} = 0.41\overline{6}$
$\frac{7}{12} = 0.58\overline{3}$	$\frac{2}{3} = 0.\overline{6}$	$\frac{5}{6} = 0.8\overline{3}$	$\frac{11}{12} = 0.91\overline{6}$

Table 5-1 Common repeating decimals

The numbers that include the fractions and their equivalent terminating or repeating decimals are called **rational numbers**. Remember that all whole numbers and integers can be written as fractions, so these numbers are also rational numbers.

Fractions and Decimals

Sometimes a problem contains both fractions and decimals.

To solve a problem with a fraction and a decimal, either change the fraction to a decimal and then perform the operation or change the decimal to a fraction and then perform the operation.

EXAMPLE
Add $\frac{3}{4} + 0.8$.

SOLUTION

$$\frac{3}{4} = 0.75 \qquad \left(\text{change } \frac{3}{4} \text{ to } 0.75\right)$$

$$+\underline{0.8} = \underline{0.80}$$
$$1.55$$

Hence, $\frac{3}{4} + 0.8 = 1.55$.

ALTERNATE SOLUTION

$$\frac{3}{4} = \frac{3}{4} = \frac{15}{20} \qquad \left(\text{change } 0.8 \text{ to } \frac{8}{10}\right)$$

$$+\underline{0.8} = \underline{\frac{8}{10}} = \underline{\frac{16}{20}}$$

$$\frac{31}{20} = 1\frac{11}{20}$$

Hence, $\frac{3}{4} + 0.8 = 1\frac{11}{20}$.

EXAMPLE
Multiply $\frac{9}{10} \times 0.75$.

SOLUTION

$$\frac{9}{10} = 0.9$$

$$0.9 \times 0.75 = 0.675$$

ALTERNATE SOLUTION

$$0.75 = \frac{3}{4}$$

$$\frac{9}{10} \times \frac{3}{4} = \frac{27}{40}$$

Hence, $\frac{9}{10} \times 0.75 = 0.675$ or $\frac{27}{40}$.

PRACTICE
Perform the indicated operation:

1. $0.6 + \dfrac{7}{8}$

2. $0.32 \times \dfrac{17}{20}$

3. $\dfrac{7}{10} - 0.6$

4. $\dfrac{5}{8} \div 0.25$

5. $0.55 + \dfrac{7}{10}$

ANSWERS

1. 1.475 or $1\frac{19}{40}$

2. 0.272 or $\dfrac{34}{125}$

3. 0.1 or $\dfrac{1}{10}$

4. 2.5 or $2\frac{1}{2}$

5. 1.25 or $1\frac{1}{4}$

Operations with Positive and Negative Decimals

When performing operations with decimals that are positive and negative, perform the operations using the rules given in Chapter 2.

EXAMPLE
Add $-8.23 + (-0.61)$.

SOLUTION

$$
\begin{array}{r}
-8.23 \\
+ \underline{-0.61} \\
-8.84
\end{array}
$$

Recall that when two negative numbers are added, the answer is negative.

EXAMPLE
Multiply -3.6×2.01.

SOLUTION

$$
\begin{array}{r}
2.01 \\
\times \underline{3.6} \\
1206 \\
\underline{603} \\
7.236
\end{array}
$$

Since a negative number times a positive number is negative, the answer -7.236.

PRACTICE
Perform the indicated operation:

1. $-18.62 + 3.25$
2. $-0.06 - 1.4$
3. $-3.7 \times (-2.95)$

4. $-18.27 - (-3.5)$
5. $4.5 - 10.95$

ANSWERS
1. -15.37
2. -1.46
3. 10.915
4. -14.77
5. -6.45

Quiz

1. In the number 2.637945, the place value of the 9 is:
   (a) hundredths
   (b) thousandths
   (c) ten thousandths
   (d) hundred thousandths

2. Name the number 0.0035.
   (a) thirty-five hundredths
   (b) thirty-five thousandths
   (c) thirty-five ten thousandths
   (d) thirty-five hundred thousandths

3. Round 0.71345 to the nearest thousandth.
   (a) 0.71
   (b) 0.713
   (c) 0.714
   (d) 0.7136

4. Add $3.85 + 2.6 + 4.032$.
   (a) 9.615
   (b) 11.035
   (c) 9.513
   (d) 10.482

5. Subtract $42.6 - 17.259$.
   (a) 26.814
   (b) 25.341
   (c) 24.431
   (d) 25.431

6. Multiply 5.6×0.024.
   (a) 0.1344
   (b) 0.1286
   (c) 0.1592
   (d) 0.3714

7. Multiply 0.002×0.009.
   (a) 0.000018
   (b) 0.0018
   (c) 0.00018
   (d) 0.0000018

8. Divide $56.852 \div 9.32$.
   (a) 6.37
   (b) 6.114
   (c) 6.1
   (d) 6.423

9. Divide $249.1 \div 47$.
   (a) 5.3
   (b) 0.53
   (c) 53
   (d) 0.053

10. Arrange in order of smallest to largest: 0.3, 0.33, 0.003, 0.303.
    (a) 0.33, 0.003, 0.3, 0.303
    (b) 0.303, 0.33, 0.3, 0.303
    (c) 0.303, 0.33, 0.003, 0.33
    (d) 0.003, 0.3, 0.303, 0.33

11. A person cut three pieces of ribbon from a piece 36 inches long. The lengths were 8.5 in., 4.75 in., and 5.3 in. How much ribbon was left?
    (a) 18.55 in.
    (b) 25.95 in.
    (c) 17.45 in.
    (d) 23.05 in.

12. A phone card charges $0.99 for the first 10 minutes and $0.08 for each minute after that. Find the cost of a 23-minute call.
    (a) $2.03
    (b) $1.84
    (c) $3.07
    (d) $1.79

13. Change $\frac{5}{16}$ to a decimal.

    (a) 3.2

    (b) 0.625

    (c) 0.3125

    (d) 0.475

14. Change $\frac{7}{11}$ to a decimal.

    (a) $0.6\overline{3}$

    (b) 0.636

    (c) $0.\overline{63}$

    (d) 0.63

15. Change 0.45 to a reduced fraction.

    (a) $\dfrac{5}{11}$

    (b) $\dfrac{45}{10}$

    (c) $\dfrac{9}{20}$

    (d) $\dfrac{45}{100}$

16. Change 0.125 to a reduced fraction.

    (a) $\dfrac{125}{1000}$

    (b) $\dfrac{1}{8}$

    (c) $\dfrac{5}{4}$

    (d) $\dfrac{3}{8}$

17. Add $\frac{3}{4} + 0.625$.

    (a) 1.375

    (b) 1.875

    (c) 0.46875

    (d) 0.575

18. Multiply $\frac{5}{6} \times 0.3$.

    (a) $\dfrac{1}{4}$

    (b) 0.75

    (c) $\dfrac{15}{6}$

    (d) 2.5

19. Subtract $1.237 - (-0.56)$.
    (a) 0.677
    (b) 1.797
    (c) -1.797
    (d) -0.677

20. Divide $-0.56 \div 8$.
    (a) -0.7
    (b) -0.07
    (c) 0.7
    (d) 7

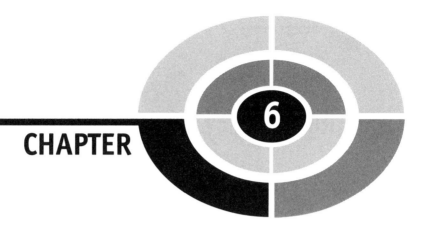

CHAPTER 6

Percent

Basic Concepts

Percent means hundredths. For example, 17% means $\frac{17}{100}$ or 0.17. Another way to think of 17% is to think of 17 equal parts out of 100 equal parts (see Fig. 6-1).

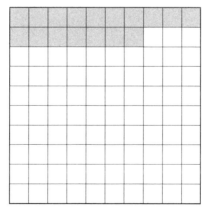

Fig. 6-1.

It should be pointed out that 100% means $\frac{100}{100}$ or 1.

Changing Percents to Decimals

To change a percent to a decimal, drop the percent sign and move the decimal point two places to the left.

EXAMPLE
Write 8% as a decimal.

SOLUTION

$$8\% = 0.08$$

Math Note: If there is no decimal point in the percent, it is at the end of the number: i.e., 8% = 8.0%.

EXAMPLE
Write 63% as a decimal.

SOLUTION

$$63\% = 0.63$$

EXAMPLE
Write 253% as a decimal.

SOLUTION

$$253\% = 2.53$$

EXAMPLE
Write 43.5% as a decimal.

SOLUTION

$$43.5\% = 0.435$$

EXAMPLE
Write 0.2% as a decimal.

SOLUTION

$$0.2\% = 0.002$$

PRACTICE

Write each of the following percent values as a decimal:

1. 26%
2. 3%
3. 125%
4. 16.6%
5. 0.8%

ANSWERS

1. 0.26
2. 0.03
3. 1.25
4. 0.166
5. 0.008

Changing Decimals to Percents

To change a decimal to a percent, move the decimal point two places to the right and affix the % sign.

Math Note: If the decimal is located at the end of the number, do not write it.

EXAMPLE

Write 0.78 as a percent.

SOLUTION

$$0.78 = 78\%$$

EXAMPLE

Write 0.02 as a percent.

SOLUTION

$$0.02 = 2\%$$

EXAMPLE
Write 1.29 as a percent.

SOLUTION

$$1.29 = 129\%$$

EXAMPLE
Write 0.682 as a percent.

SOLUTION

$$0.682 = 68.2\%$$

EXAMPLE
Write 0.0051 as a percent.

SOLUTION

$$0.0051 = 0.51\%$$

EXAMPLE
Write 8 as a percent.

SOLUTION

$$8 = 8.00 = 800\%$$

PRACTICE
Write each of the following decimals as percents:

1. 0.09
2. 0.91
3. 0.375
4. 0.0042
5. 3

ANSWERS
1. 9%
2. 91%
3. 37.5%
4. 0.42%
5. 300%

Changing Fractions to Percents

To change a fraction to a percent, change the fraction to a decimal (i.e., divide the numerator by the denominator), then move the decimal two places to the right and affix the % sign.

EXAMPLE
Write $\frac{3}{4}$ as a percent.

SOLUTION

$$= 4\overline{)3.00} \quad \begin{array}{r} .75 \\ \underline{28} \\ 20 \\ \underline{20} \\ 0 \end{array}$$

$\frac{3}{4} = 75\%$

EXAMPLE
Write $\frac{2}{5}$ as a percent.

SOLUTION

$$= 5\overline{)2.0} \quad \begin{array}{r} .4 \\ \underline{20} \\ 0 \end{array}$$

$\frac{2}{5} = 40\%$

EXAMPLE
Write $\frac{3}{8}$ as a percent.

SOLUTION

$$= 8\overline{)3.000} \quad \begin{array}{r} .375 \\ \underline{24} \\ 60 \\ \underline{56} \\ 40 \\ \underline{40} \\ 0 \end{array}$$

$\frac{3}{8} = 37.5\%$

EXAMPLE
Write $1\frac{1}{4}$ as a percent.

SOLUTION
$1\frac{1}{4} = \frac{5}{4}$

$$
\begin{array}{r}
1.25 \\
4\overline{)5.00} \\
\underline{4} \\
10 \\
\underline{8} \\
20 \\
\underline{20} \\
0
\end{array}
$$

$1\frac{1}{4} = 125\%$

EXAMPLE
Write $\frac{5}{6}$ as a percent.

SOLUTION

$$
\begin{array}{r}
.833 \\
6\overline{)5.000} \\
\underline{48} \\
20 \\
\underline{18} \\
20 \\
\underline{18} \\
2
\end{array}
$$

$\frac{5}{6} = 83.\overline{3}\%$

PRACTICE
Write each of the following fractions as percents:

1. $\dfrac{1}{2}$

2. $\dfrac{7}{8}$

3. $\dfrac{9}{50}$

4. $6\frac{3}{4}$

5. $\dfrac{5}{12}$

ANSWERS

1. 50%
2. 87.5%
3. 18%
4. 675%
5. $41.\overline{6}\%$

Changing Percents to Fractions

To change a percent to a fraction, write the numeral in front of the percent sign as the numerator of a fraction whose denominator is 100. Reduce or simplify the fraction if possible.

EXAMPLE
Write 45% as a fraction.

SOLUTION

$$45\% = \frac{45}{100} = \frac{9}{20}$$

EXAMPLE
Write 80% as a fraction.

SOLUTION

$$80\% = \frac{80}{100} = \frac{4}{5}$$

EXAMPLE
Write 6% as a fraction.

SOLUTION

$$6\% = \frac{6}{100} = \frac{3}{50}$$

EXAMPLE
Write 175% as a fraction.

SOLUTION

$$175\% = \frac{175}{100} = 1\frac{75}{100} = 1\frac{3}{4}$$

PRACTICE
Write each of the following percents as fractions:

1. 2%
2. 85%
3. 145%
4. 30%
5. 97%

ANSWERS

1. $\frac{1}{50}$

2. $\frac{17}{20}$

3. $1\frac{9}{20}$

4. $\frac{3}{10}$

5. $\frac{97}{100}$

Three Types of Percent Problems

There are three basic types of percent problems, and there are several different methods that can be used to solve these problems. The circle method will be used in this chapter. The equation method will be used in the next chapter, and finally, the proportion method will be used in Chapter 8. A percent problem has three values: the base (B) or whole, the rate (R) or percent, and the part (P). For example, if a class consisted of 20 students, 5 of whom were absent today, the base or whole would be 20, the part would be 5, and the rate or percent of students who were absent would be $\frac{5}{20} = 25\%$.

Draw a circle and place a horizontal line through the center and a vertical line halfway down in the center also. In the top section, write the word "is." In the lower left section write the % sign, and in the lower right section write the word "of." In the top section place the part (P). In the lower left section place the rate (R) or percent number, and in the lower right section place the base (B). One of these three quantities will be unknown (see Fig. 6-2).

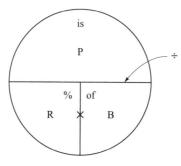

Fig. 6-2.

If you are given the two bottom numbers, multiply them to get the top number: i.e., $P = R \times B$. If you are given the top number and one of the bottom numbers, divide to find the other number: i.e., $R = \frac{P}{B}$ or $B = \frac{P}{R}$. (see Fig. 6-3).

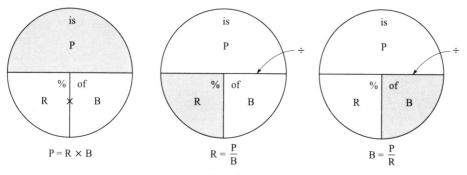

Fig. 6-3.

TYPE 1   FINDING THE PART

Type 1 problems can be stated as follows:

- "Find 20% of 60."
- "What is 20% of 60?"
- "20% of 60 is what number?"

In Type 1 problems, you are given the base and the rate and are asked to find the part. From the circle: P = R × B. Here, then, you change the percent to a decimal or fraction and multiply.

EXAMPLE
Find 40% of 80.

SOLUTION
Draw the circle and put 40% in the % section and 80 in the "of" section (see Fig. 6-4).

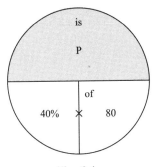

Fig. 6-4.

Change the percent to a decimal and multiply: $0.40 \times 80 = 32$.

EXAMPLE
Find 25% of 60.

SOLUTION
Draw the circle and put 25 in the % section and 60 in the "of" section (see Fig. 6-5).

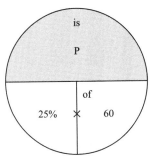

Fig. 6-5.

Change 25% to a decimal and multiply: $0.25 \times 60 = 15$.

Math Note: Always change the percent to a decimal or fraction before multiplying or dividing.

PRACTICE
1. Find 50% of 90.
2. Find 24% of 16.
3. What is 48% of 125?
4. 62.5% of 32 is what number?
5. Find 16% of 230.

ANSWERS
1. 45
2. 3.84
3. 60
4. 20
5. 36.8

TYPE 2   FINDING THE RATE

Type 2 problems can be stated as follows:

- "What percent of 16 is 10?"
- "10 is what percent of 16?"

In type 2 problems, you are given the base and the part and are asked to find the rate or percent. The formula is $R = \frac{P}{B}$. In this case, divide the part by the base and then change the answer to a percent (see Fig. 6-6).

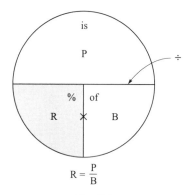

Fig. 6-6.

EXAMPLE
What percent of 5 is 2?

SOLUTION
Draw the circle and place 5 in the "of" section and 2 in the "is" section (see Fig. 6-7).

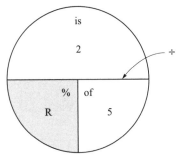

Fig. 6-7.

Then divide $\frac{2}{5} = 2 \div 5 = 0.40$ Change the decimal to a percent: $0.40 = 40\%$.

EXAMPLE
45 is what percent of 60?

SOLUTION
Draw the circle and put 45 in the "is" section and 60 in the "of" section (see Fig. 6-8).

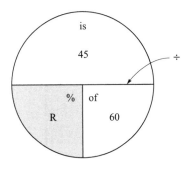

Fig. 6-8.

Then divide $\frac{45}{60} = 45 \div 60 = 0.75$. Change the decimal to a percent: $0.75 = 75\%$.

PRACTICE
1. What percent of 10 is 3?
2. 15 is what percent of 120?
3. 8 is what percent of 40?
4. What percent of 80 is 32?
5. What percent of 75 is 50?

ANSWERS
1. 30%
2. 12.5%
3. 20%
4. 40%
5. $66.\overline{6}\%$

TYPE 3   FINDING THE BASE

Type 3 problems can be stated as follows:

- "16 is 20% of what number?"
- "20% of what number is 16?"

In type 3 problems, you are given the rate and the part, and you are asked to find the base. From the circle: $B = \frac{P}{R}$ (see Fig. 6-9).

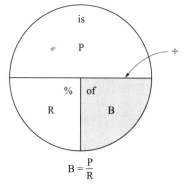

$$B = \frac{P}{R}$$

Fig. 6-9.

EXAMPLE
42% of what number is 294?

SOLUTION
Draw the circle and place 42 in the percent section and 294 in the "is" section (see Fig. 6-10).

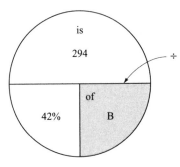

Fig. 6-10.

Change 42% to 0.42 and divide: $294 \div 0.42 = 700$.

EXAMPLE
36 is 80% of what number?

SOLUTION
Draw the circle and place 36 in the "is" section and 80 in the percent section (see Fig. 6-11).

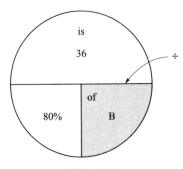

Fig. 6-11.

Change 80% to 0.80 and divide: $36 \div 0.80 = 45$.

PRACTICE
1. 2% of what number is 46?
2. 150 is 50% of what number?
3. 42 is 60% of what number?
4. 30% of what number is 27?
5. 18.6% of what number is 139.5?

ANSWERS
1. 2300
2. 300
3. 70
4. 90
5. 750

Word Problems

Percent word problems can be done using the circle. In the lower left section place the *rate* or *percent*. In the lower right section place the *base* or *whole*, and in the upper section, place the *part* (see Fig. 6-12).

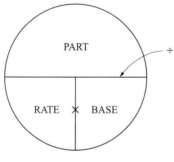

Fig. 6-12.

In order to solve a percent problem:

- *Step 1   Read the problem.*
- *Step 2   Identify the base, rate (%), and part. One will be unknown.*
- *Step 3   Substitute the values in the circle.*
- *Step 4   Perform the correct operation: i.e., either multiply or divide.*

EXAMPLE
On a test consisting of 40 problems, a student received a grade of 95%. How many problems did the student answer correctly?

SOLUTION
Place the rate, 95%, and the base, 40, into the circle (see Fig. 6-13).

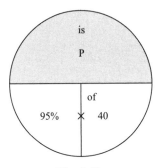

Fig. 6-13.

Change 95% to 0.95 and multiply $0.95 \times 40 = 38$. Hence, the student got 38 problems correct.

EXAMPLE
A football team won 9 of its 12 games. What percent of the games played did the team win?

SOLUTION
Place the 9 in the "part" section and the 12 in the "base" section of the circle (see Fig. 6-14).

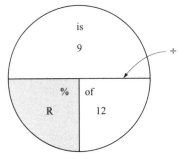

Fig. 6-14.

Then divide $\frac{9}{12} = 9 \div 12 = 0.75$. Hence, the team won 75% of its games.

EXAMPLE
The sales tax rate in a certain state is 6%. If sales tax on an automobile was $1,110, find the price of the automobile.

SOLUTION
Place the 6% in the "rate" section and the $1,110 in the "part" section (see Fig. 6-15).

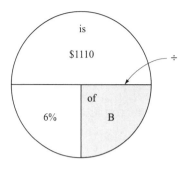

Fig. 6-15.

Change the 6% to 0.06 and divide: $1,110 ÷ 0.06 = $18,500. Hence, the price of the automobile was $18,500.

Another type of percent problem you will often see is the percent increase or percent decrease problem. In this situation, always remember that the *old* or *original* number is used as the base.

EXAMPLE
A coat that originally cost $150 was reduced to $120. What was the percent of the reduction?

SOLUTION
Find the amount of reduction: $150 − $120 = $30. Then place $30 in the "part" section and $150 in the "base" section of the circle. $150 is the base since it was the *original* price (see Fig. 6-16).

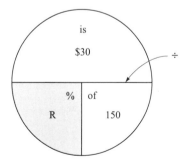

Fig. 6-16.

Divide $\frac{30}{150} = 30 ÷ 150 = 0.20$. Hence, the cost was reduced 20%.

PRACTICE
1. A person bought an item at a 15% off sale. The discount was $90. What was the regular price?

2. In an English Composition course which has 60 students, 15% of the students are mathematics majors. How many students are math majors?

3. A student correctly answered 35 of the 40 questions on a mathematics exam. What percent did she answer correctly?

4. A car dealer bought a car for $1500 at an auction and then sold it for $3500. What was the percent gain on the cost?

5. An automobile service station inspected 215 vehicles, and 80% of them passed. How many vehicles passed the inspection?

6. The sales tax rate in a certain state is 7%. How much tax would be charged on a purchase of $117.50?

7. A person saves $200 a month. If her annual income is $32,000 per year, what percent of her income is she saving?

8. A house sold for $80,000. If the salesperson receives a 4% commission, what was the amount of his commission?

9. If the markup on a microwave oven is $42 and the oven sold for $98, find the percent of the markup on the selling price.

10. A video game which originally sold for $40 was reduced $5. Find the percent of reduction.

ANSWERS
1. $600
2. 9
3. 87.5%
4. $133.\overline{3}$%
5. 172
6. $8.23
7. 7.5%
8. $3200
9. 42.86%
10. 12.5%

Quiz

1. Write 6% as a decimal.
   (a) 0.006
   (b) 0.06
   (c) 0.6
   (d) 6.0

2. Write 54.8% as a decimal.
   (a) 0.548
   (b) 5.48
   (c) 0.0548
   (d) 5480

3. Write 175% as a decimal.
   (a) 0.175
   (b) 1.75
   (c) 17.5
   (d) 1750

4. Write 0.39 as a percent.
   (a) 0.39%
   (b) 39%
   (c) 0.039%
   (d) 390%

5. Write 0.511 as a percent.
   (a) 0.511%
   (b) 0.00511%
   (c) 51.1%
   (d) 511%

6. Write 9 as a percent.
   (a) 9%
   (b) 90%
   (c) 900%
   (d) 0.09%

7. Write $\frac{7}{10}$ as a percent.

   (a) 7%
   (b) 700%
   (c) 70%
   (d) 0.7%

8. Write $\frac{2}{3}$ as a percent.

   (a) 1.5%
   (b) $23.\overline{3}$%
   (c) $32.\overline{2}$%
   (d) $66.\overline{6}$%

9. Write $2\frac{7}{8}$ as a percent.

(a) 2.875%
(b) 28.75%
(c) 287.5%
(d) 2875%

10. Write 72% as a fraction.

(a) $\dfrac{7}{25}$

(b) $7\frac{2}{10}$

(c) $\dfrac{1}{72}$

(d) $\dfrac{18}{25}$

11. Write 5% as a fraction.

(a) $\dfrac{1}{20}$

(b) $\dfrac{5}{12}$

(c) $\dfrac{3}{5}$

(d) $\dfrac{1}{12}$

12. Write 125% as a fraction.

(a) $1\frac{5}{8}$
(b) $1\frac{1}{4}$
(c) $1\frac{3}{4}$
(d) $1\frac{2}{5}$

13. Find 25% of 180.
(a) 0.45
(b) 4.5
(c) 45
(d) 4,500

14. 6 is what percent of 25?
(a) 150%

(b) 24%

(c) 41.$\overline{6}$%

(d) 72%

15. 3% of what number is 90?

(a) 2.7

(b) 270

(c) 300

(d) 3000

16. 72% of 360 is what number?

(a) 25.920

(b) 50

(c) 259.2

(d) 500

17. A person bought a house for $92,000 and made a 20% down payment. How much was the down payment?

(a) $1,840

(b) $18,400

(c) $4,600

(d) $460

18. If the sales tax rate is 5% and the sales tax on an item is $1.65, find the cost of the item.

(a) $3

(b) $33

(c) $0.08

(d) $80

19. A person earned a commission of $285 on an item that sold for $1,900. Find the rate.

(a) 5%

(b) 15%

(c) 150%

(d) 6%

20. A salesperson sold 4 chairs for $75 each and a table for $200. If the commission rate is 9.5%, find the person's commission.

(a) $4.75

(b) $26.13

(c) $475.00

(d) $47.50

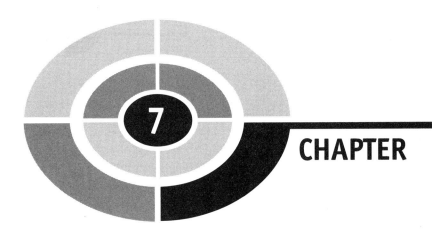

CHAPTER

7

Expression and Equations

Basic Concepts

In algebra, letters are used as variables. A **variable** can assume values of numbers. Numbers are called **constants**.

> **Math Note:** In some cases, a letter may represent a specific constant. As you will see in Chapter 9, the Greek letter pi (π) represents a constant.

An **algebraic expression** consists of variables, constants, operation signs, and grouping symbols. In the algebraic expression 3x, the "3" is a constant and the "x" is a variable. When no sign is written between a number and a variable or between two or more variables, it means multiplication. Hence the expression "3x" means "3 times x or to multiply 3 by the value of x. The expression abc means a times b times c or a × b × c.

The number before the variable is called the **numerical coefficient**. In the algebraic expression "3x", the 3 is the numerical coefficient. When the numerical coefficient is 1, it is usually not written and vice versa. Hence, xy means 1xy. Likewise, 1xy is usually written as xy. Also −xy means −1xy.

An algebraic expression consists of one or more **terms**. A **term** is a number or variable, or a product or a quotient of numbers and variables. Terms are connected by + or − signs. For example, the expression 3x + 2y − 6 has 3 terms. The expression 8p + 2q has two terms, and the expression $6x^2y$ consists of one term.

Evaluating Algebraic Expressions

*In order to **evaluate** an algebraic expression, substitute the values of the variables in the expression and simplify using the order of operations.*

EXAMPLE
Evaluate 2xy when x = 2 and y = 5.

SOLUTION
$$2xy = 2(2)(5)$$
$$= 20$$

Hence, when x = 2 and y = 5, the value of the expression 2xy is 20.

EXAMPLE
Evaluate $-3x^2$ when x = −6

SOLUTION
$$-3x^2 = -3(-6)^2$$
$$= -3(36)$$
$$= -108$$

EXAMPLE
Evaluate 5(x + 3y) when x = −5 and y = 3.

SOLUTION

$$5(x + 3y) = 5(-5 + 3(3))$$
$$= 5(-5 + 9)$$
$$= 5(4)$$
$$= 20$$

PRACTICE
Evaluate each of the following expressions:

1. 17 − x when x = 7
2. x + 2y when x = −5 and y = 8
3. $(2x + 5)^2$ when x = 6
4. 3x − 2y + z when x = −3, y = 4, and z = 3
5. $5x^2 - 2y^2$ when x = 3 and y = 6

ANSWERS
1. 10
2. 11
3. 289
4. −14
5. −27

The Distributive Property

An important property that is often used in algebra is called the **distributive property**:

For any numbers a, b, and c, a(b + c) = ab + ac.

The distributive property states that when a sum of two numbers or expressions in parentheses is multiplied by a number outside the parentheses, the same result will occur if you multiply each number or expression inside the parentheses by the number outside the parentheses.

EXAMPLE
Multiply 5(2x + 3y).

SOLUTION

$$5(2x + 3y) = 5 \cdot 2x + 5 \cdot 3y$$
$$= 10x + 15y$$

EXAMPLE
Multiply $8(x + 3y)$.

SOLUTION

$$8(x + 3y) = 8 \cdot 1x + 8 \cdot 3y$$
$$= 8x + 24y$$

The distributive property also works for subtraction.

EXAMPLE
Multiply $9(2x - 4y)$.

SOLUTION

$$9(2x - 4y) = 9 \cdot 2x - 9 \cdot 4y$$
$$= 18x - 36y$$

The distributive property works for the sum or difference of three or more expressions in parentheses.

EXAMPLE
Multiply $3(6a + 2b - 7c)$.

SOLUTION

$$3(6a + 2b - 7c) = 3 \cdot 6a + 3 \cdot 2b - 3 \cdot 7c$$
$$= 18a + 6b - 21c$$

When a negative number is multiplied, make sure to change the signs of the terms inside the parentheses when multiplying.

EXAMPLE
Multiply $-2(6p - 7q - 3r)$.

SOLUTION

$$-2(6p - 7q - 3r) = -2 \cdot 6p - 2(-7q) - 2(-3r)$$
$$= -12p + 14q + 6r$$

> **Math Note:** The distributive property is called the distributive property for multiplication over addition.

PRACTICE
Multiply each of the following:

1. $6(3x + 8y)$
2. $2(b + 4)$
3. $4(7a - 2b + 3c)$
4. $-3(2p - 5q + 3r)$
5. $-5(6x - 7y + z)$

ANSWERS
1. $18x + 48y$
2. $2b + 8$
3. $28a - 8b + 12c$
4. $-6p + 15q - 9r$
5. $-30x + 35y - 5z$

Combining Like Terms

Like terms have the same variables and the same exponents of the variables. For example, 3x and $-5x$ are like terms because they have the same variables, whereas 5x and $-8y$ are **unlike terms** because they have different variables. Table 7-1 will help you to identify like terms.

Like terms	Unlike terms
$2x, -7x$	$2x, -7y$
$-8x^2, 3x^2$	$-8x^2, 3x$
$13xy, -7xy$	$13xy, -7xz$
$5x^2y, 3x^2y$	$5x^2y, 3xy^2$
$x, 4x$	$x, 4$

Table 7-1 Identification of like terms

Like terms can be added or subtracted: e.g., $4x + 2x = 6x$.

To add or subtract like terms, add or subtract the numerical coefficients and use the same variables.

EXAMPLE
Add 7y + 3y.

SOLUTION

$$7y + 3y = (7 + 3)y$$
$$= 10y$$

EXAMPLE
Subtract 8x − 5x.

SOLUTION

$$8x - 5x = (8 - 5)x$$
$$= 3x$$

Adding or subtracting like terms is often called "combining like terms."

EXAMPLE
Combine like terms: 9x + 7x − 3x − 2x.

SOLUTION

$$9x + 7x - 3x - 2x = (9 + 7 - 3 - 2)x.$$
$$= 11x$$

Unlike terms cannot be added or subtracted. When an expression has both like and unlike terms, it can be simplified by combining all like terms.

EXAMPLE
Combine like terms: 6x − 8x + 2y − 3y + 5x − 4y.

SOLUTION

$$6x - 8x + 2y - 3y + 5x - 4y = 6x - 8x + 5x + 2y - 3y - 4y$$
$$= (6 - 8 + 5)x + (2 - 3 - 4)y$$
$$= 3x + (-5)y$$
$$= 3x - 5y$$

> **Math Note:** An algebraic expression of the form a +(−b) is usually written as a − b.

PRACTICE

For each of the following expressions, combine like terms:

1. $10x + 21x$
2. $18y − 12y$
3. $x − 6x$
4. $3x + 5x + 6x − 2x$
5. $−4y − 7y − 11y$
6. $25x − 2y + 13x − 6y + x$
7. $4x + 2y − 6 + 3x − 5y +2$
8. $7a − 6b + 3c − 4a − 2b + 5c$
9. $x + 2y − x − y + 5x$
10. $8p + q − 6p − q$

ANSWERS

1. $31x$
2. $6y$
3. $−5x$
4. $12x$
5. $−22y$
6. $39x − 8y$
7. $7x − 3y − 4$
8. $3a − 8b + 8c$
9. $5x + y$
10. $2p$

Removing Parentheses and Combining Like Terms

Sometimes in algebra it is necessary to use the distributive property to remove parentheses and then combine terms. This procedure is also called "simplifying" an algebraic expression.

EXAMPLE
Simplify $5(2x + 3) - 6$.

SOLUTION

$$5(2x + 3) - 6 = 10x + 15 - 6 \qquad \text{Remove parentheses}$$
$$= 10x + 9 \qquad \text{Combine like terms}$$

EXAMPLE
Simplify $-2(x + 7y) + 6x$.

SOLUTION

$$-2(x + 7y) + 6x = -2x - 14y + 6x \qquad \text{Remove parentheses}$$
$$= 4x - 14y \qquad \text{Combine like terms}$$

PRACTICE
Simplify each of the following expressions:

1. $7(x + 2) + 8$
2. $3(2x + 3y - 5) + 6x - 3$
3. $-2(4a + 6b) - 2a + 3b$
4. $5(a + 2b - c) + 6a$
5. $9(x + 2y) + 4(x - 6y)$

ANSWERS
1. $7x + 22$
2. $12x + 9y - 18$
3. $-10a - 9b$
4. $11a + 10b - 5c$
5. $13x - 6y$

Formulas

In mathematics and science, many problems can be solved using *formulas*. A **formula** is a mathematical statement of a relationship of two or more variables. For example, the relationship between Fahrenheit temperature and Celsius temperature can be expressed by the formula $F = \frac{9}{5}C + 32°$. Formulas can be evaluated in the same way as expressions are evaluated:

i.e., substitute the values of the variables in the formula and simplify using the order of operations.

EXAMPLE
Find the Fahrenheit temperature corresponding to a Celsius temperature of $-10°$. Use the formula $F = \frac{9}{5}C + 32°$.

SOLUTION

$$F = \frac{9}{5}C + 32°$$
$$= \frac{9}{5}(-10) + 32$$
$$= 9(-2) + 32$$
$$= -18 + 32$$
$$= 14$$

Hence, a Celsius temperature of $-10°$ is equivalent to a Fahrenheit temperature of $14°$.

EXAMPLE
Find the interest on a loan whose principal (P) is \$3,000 with a rate (R) of 6% for 3 years (T). Use $I = PRT$.

SOLUTION

$$I = PRT$$
$$= (\$3,000)(0.06)(3)$$
$$= \$540$$

PRACTICE
1. Find the distance (D) in miles an automobile travels in 3 hours (T) at a rate (R) of 40 miles per hour. Use $D = RT$.
2. Find the Celsius (C) temperature when the Fahrenheit (F) temperature is $77°$. Use $C = \frac{5}{9}(F - 32°)$.
3. Find the volume (V) in cubic inches of a cube when the length of a side (s) is 6 inches. Use $V = s^3$.
4. Find the perimeter (P) (distance around the outside) of a rectangle if the length (l) is 6 feet and the width (w) is 4 feet. Use $P = 2l + 2w$.
5. Find the current (I) in amperes when the electromotive force (E) is 12 volts and the resistance (R) is 3 ohms. Use $I = \frac{E}{R}$.

ANSWERS
1. 120 miles
2. 25°
3. 216 cubic inches
4. 20 feet
5. 4 amperes

Solving Simple Equations

An **equation** is a statement of equality of two algebraic expressions: e.g., $3 + 2 = 5$ is an equation. An equation can contain one or more variables: e.g., $x + 7 = 10$ is an equation with one variable, x. If an equation has no variables, it is called a **closed** equation. Closed equations can be either true or false: e.g., $6 + 8 = 14$ is a closed true equation, whereas $3 + 5 = 6$ is a false closed equation. Open equations, also called **conditional** equations, are neither true nor false. However, if a value for the variable is substituted in the equation and a closed true equation results, the value is called a **solution** or **root** of the equation. For example, when 3 is substituted for x in the equation $x + 7 = 10$, the resulting equation $3 + 7 = 10$ is true, so 3 is called a solution of the equation. Finding the solution of an equation is called **solving** the equation. The expression to the left of the equal sign in an equation is called the **left member** or **left side** of the equation. The expression to the right of the equal sign is called the **right member** or **right side** of the equation.

In order to solve an equation, it is necessary to transform the equation into a simpler equivalent equation with only the variable on one side and a constant on the other side. There are four basic types of equations and four principles that are used to solve them. These principles do not change the nature of an equation: i.e., the simpler equivalent equation has the same solution as the original equation.

In order to **check** an equation, substitute the value of the solution or root for the variable in the original equation and see if a closed true equation results.

An equation such as $x - 8 = 22$ can be solved by using the **addition principle**. *The same number can be added to both sides of an equation without changing the nature of the equation.*

EXAMPLE
Solve $x - 8 = 22$.

SOLUTION

$$x - 8 = 22$$
$$x - 8 + 8 = 22 + 8 \qquad \text{Add 8 to both sides}$$
$$x - 0 = 30$$
$$x = 30$$

Check:

$$x - 8 = 22$$
$$30 - 8 = 22$$
$$22 = 22$$

An equation such as $x + 5 = 9$ can be solved by the **subtraction principle**. The same number can be subtracted from both members of the equation without changing the nature of the equation.

EXAMPLE

Solve $x + 5 = 9$.

SOLUTION

$$x + 5 = 9$$
$$x + 5 - 5 = 9 - 5 \qquad \text{Subtract 5 from both sides}$$
$$x + 0 = 4$$
$$x = 4$$

Check:

$$x + 5 = 9$$
$$4 + 5 = 9$$
$$9 = 9$$

An equation such as $7x = 28$ can be solved by using the **division principle.** *Both sides of an equation can be divided by the same non-zero number without changing the nature of the equation.*

EXAMPLE

Solve $7x = 28$.

SOLUTION

$$7x = 28$$

$$\frac{7x}{7} = \frac{28}{7} \qquad \text{Divide both sides by 7}$$

$$x = 4$$

Check:

$$7x = 28$$

$$7(4) = 28$$

$$28 = 28$$

An equation such as $\frac{x}{3} = 15$ can be solved by using the **multiplication principle.** *Both sides of an equation can be multiplied by the same non-zero number without changing the nature of the equation.*

EXAMPLE

Solve $\frac{x}{3} = 15$.

SOLUTION

$$\frac{x}{3} = 15$$

$$\overset{1}{\underset{1}{\frac{3}{1}}} \cdot \frac{x}{\underset{1}{3}} = 15 \cdot 3$$

$$x = 45$$

Check:

$$\frac{x}{3} = 15$$

$$\frac{45}{3} = 15$$

$$15 = 15$$

As you can see, there are four basic types of equations and four basic principles that are used to solve them. Before attempting to solve an equation, you should see what operation is being performed on the variable and then use the opposite principle to solve the equation. Addition and subtraction are opposite operations and multiplication and division are opposite operations.

PRACTICE

Solve each of the following equations:

1. $x + 40 = 62$
2. $x - 19 = 32$
3. $8x = 56$
4. $23 + x = 6$
5. $x - 6 = -14$
6. $\dfrac{x}{5} = 2$
7. $109 = x + 78$
8. $-5x = 35$
9. $x + 8 = -10$
10. $\dfrac{x}{8} = 3$

ANSWERS

1. 22
2. 51
3. 7
4. −17
5. −8
6. 10
7. 31
8. −7
9. −18
10. 24

Solving Equations Using Two Principles

Most equations require you to use more than one principle to solve them. These equations use the addition or subtraction principle first and then use the division principle.

EXAMPLE

Solve $8x + 5 = 37$.

SOLUTION

$$8x + 5 = 37$$
$$8x + 5 - 5 = 37 - 5 \qquad \text{Subtract 5}$$
$$8x = 32$$

$$\frac{\cancel{8}x}{\cancel{8}} = \frac{32}{8}$$

$$x = 4$$

Check:

$$8x + 5 = 37$$
$$8(4) + 5 = 37$$
$$32 + 5 = 37$$
$$37 = 37$$

EXAMPLE
Solve $4x + 20 = -4$.

SOLUTION

$$4x + 20 = -4$$
$$x + 20 - 20 = -4 - 20 \qquad \text{Subtract 20}$$
$$4x = -24$$
$$\frac{\cancel{4}x}{\cancel{4}} = \frac{-24}{4} \qquad \text{Divide by 4}$$
$$x = -6$$

Check:

$$4x + 20 = -4$$
$$4(-6) + 20 = -4$$
$$-24 + 20 = -4$$
$$-4 = -4$$

EXAMPLE
Solve $-3x + 12 = -36$.

SOLUTION

$$-3x + 12 = -36$$

$$-3x + 12 - 12 = -36 - 12 \qquad \text{Subtract 12}$$

$$-3x = -48$$

$$\frac{-3x}{-3} = \frac{-48}{-3} \qquad \text{Divide by } -3$$

$$x = 16$$

Check:

$$-3x + 12 = -36$$

$$-3(16) + 12 = -36$$

$$-48 + 12 = -36$$

$$-36 = -36$$

PRACTICE

Solve each of the following equations:

1. $7x + 8 = 71$
2. $5 + 9x = 50$
3. $6x - 3 = 21$
4. $-2x + 10 = 16$
5. $45 - 5x = 50$

ANSWERS

1. 9
2. 5
3. 4
4. −3
5. −1

Solving More Difficult Equations

When an equation has like terms on the same side, these terms can be combined first.

EXAMPLE

Solve $2x + 5 + 3x = 30$.

SOLUTION

$$2x + 5 + 3x = 30$$
$$5x + 5 = 30 \qquad \text{Combine } 2x + 3x$$
$$5x + 5 - 5 = 30 - 5 \qquad \text{Subtract 5}$$
$$5x = 25$$
$$\frac{5x}{5} = \frac{25}{5} \qquad \text{Divide by 5}$$
$$x = 5$$

Check:

$$2x + 5 + 3x = 30$$
$$2(5) + 5 + 3(5) = 30$$
$$10 + 5 + 15 = 30$$
$$15 + 15 = 30$$
$$30 = 30$$

If an equation has parentheses, use the distributive property to remove parentheses.

EXAMPLE
Solve $3(2x + 6) = 30$.

SOLUTION

$$3(2x + 6) = 30$$
$$6x + 18 = 30 \qquad \text{Remove parentheses}$$
$$6x + 18 - 18 = 30 - 18 \qquad \text{Subtract 18}$$
$$6x = 12$$
$$\frac{6x}{6} = \frac{12}{6} \qquad \text{Divide by 6}$$
$$x = 2$$

Check:

$$3(2x + 6) = 30$$
$$3(2 \cdot 2 + 6) = 30$$
$$3 \cdot 10 = 30$$
$$30 = 30$$

The procedure for solving equations in general is:

- *Step 1   Remove parentheses.*
- *Step 2   Combine like terms on each side of the equation.*
- *Step 3   Use the addition and/or subtraction principle to get the variables on one side and the constant terms on the other side.*
- *Step 4   Use the division principle to solve for x.*
- *Step 5   Check the equation.*

EXAMPLE

Solve $5(2x - 7) - 3x = 5x + 9$.

SOLUTION

$$5(2x - 7) - 3x = 5x + 9$$

$$10x - 35 - 3x = 5x + 9 \qquad \text{Remove parentheses}$$

$$7x - 35 = 5x + 9 \qquad \text{Combine like terms}$$

$$7x - 35 - 5x = 5x - 5x + 9 \qquad \text{Get variables on one side and the}$$

$$2x - 35 = 9 \qquad \text{constants on the other side}$$

$$2x - 35 + 35 = 9 + 35$$

$$2x = 44$$

$$\frac{2x}{2} = \frac{44}{2} \qquad \text{Divide by 2}$$

$$x = 22$$

Check:

$$5(2x - 7) - 3x = 5x + 9$$

$$5(2 \cdot 22 - 7) - 3 \cdot 22 = 5 \cdot 22 + 9$$

$$5(44 - 7) - 66 = 110 + 9$$

$$5(37) - 66 = 119$$

$$185 - 66 = 119$$

$$119 = 119$$

PRACTICE

Solve each of the following equations:

1. $6x + 4 = 2x + 12$
2. $x - 10 = 32 - 5x$
3. $9x - 5 + 2x = 17$
4. $7x - 3 = 4x - 27$
5. $10x + 36 = 6x$
6. $3(x - 2) = 24$
7. $4(2x - 5) = 20$
8. $8x - 2(3x - 7) = 20$
9. $5(x + 6) = 10$
10. $9(3x - 1) = 8(3x - 6)$

ANSWERS

1. 2
2. 7
3. 2
4. −8
5. −9
6. 10
7. 5
8. 3
9. −4
10. −13

Algebraic Representation of Statements

In order to solve word problems in algebra, it is necessary to translate the verbal statements into algebraic expressions. An unknown can be designated by a variable, usually x. For example, the statement "four times a number plus ten" can be written algebraically as "$4x + 10$."

EXAMPLE

Write each of the following statements in symbols:

1. the sum of a number and 15
2. seven subtracted from two times a number
3. the product of six and a number
4. three times the sum of a number and 5
5. a number divided by four

SOLUTION

1. $x + 15$
2. $2x - 7$
3. $6x$
4. $3(x + 5)$
5. $x \div 4$

In solving word problems it is also necessary to represent two quantities using the same variable. For example, if the sum of two numbers is 10, and one number is x, the other number would be $10 - x$. The reason is that given one number, say 7, you can find the other number by subtracting $10 - 7$ to get 3. Another example: suppose you are given two numbers and the condition that one number is twice as large as the other. How would you represent the two numbers? The smaller number would be x, and since the second number is twice as large, it can be represented as 2x.

EXAMPLE

Represent algebraically two numbers such that one number is three more than twice the other number.

SOLUTION

Let x = one number
Let $2x + 3$ = the other number

EXAMPLE

Represent algebraically two numbers such that the difference between the two numbers is 7.

SOLUTION

Let x = one number
Let $x - 7$ = the other number

PRACTICE

1. Represent algebraically two numbers so that one number is 5 more than another number.
2. Represent algebraically two numbers so that their sum is 50.
3. Represent algebraically two numbers such that one number is five times larger than another number.
4. Represent algebraically two numbers such that one number is half another number.
5. Represent algebraically two numbers such that one number is 10 more than twice the other number.

ANSWERS
1. x, x + 5
2. x, 50 − x
3. x, 5x
4. x, $\frac{1}{2}$x or x, 2x
5. x, 2x + 10

Word Problems

Equations can be used to solve word problems in algebra.
After reading the problem:

- *Step 1   Represent one of the unknown quantities as x and the other unknown by an algebraic expression in terms of x.*
- *Step 2   Write an equation using the unknown quantities.*
- *Step 3   Solve the equation for x and then find the other unknown.*
- *Step 4   Check the answers.*

EXAMPLE
One number is 8 more than another number. The sum of the two numbers is 114. Find the numbers.

SOLUTION
Step 1   Let x = the smaller number
        Let x + 8 = the larger number
Step 2   x + x + 8 = 114
Step 3   x + x + 8 = 114

$$2x + 8 = 114$$
$$2x + 8 - 8 = 114 - 8$$
$$2x = 106$$
$$\frac{2x}{2} = \frac{106}{2}$$
$$x = 53$$
$$x + 8 = 53 + 8 = 61$$

Hence, one number is 53 and the other number is 61.
Step 4   Check: 53 + 61 = 114.

EXAMPLE

A carpenter cuts a board 72 inches long into two pieces. One piece is twice as long as the other. Find the lengths of the pieces.

SOLUTION

Step 1   Let x = the length of the smaller piece

             Let 2x = the length of the larger piece

Step 2   x + 2x = 72

Step 3   x + 2x = 72

$$3x = 72$$

$$\frac{\cancel{3}x}{\cancel{3}} = \frac{72}{3}$$

$$x = 24 \text{ inches}$$

$$2x = 2 \cdot 24 = 48 \text{ inches}$$

Hence, one piece is 24 inches long and the other piece is 48 inches long.

Step 4   Check: 24 inches + 48 inches = 72 inches

EXAMPLE

$40 in tips is to be divided among three food servers. The first server gets twice as much as the second server since he worked twice as long; the third server gets $7. Find the amount each server received.

SOLUTION

Step 1   Let x = the amount the second server receives

             Let 2x = the amount the first server received

Step 2   x + 2x + 7 = 40

$$3x + 7 = 40$$

$$3x + 7 - 7 = 40 - 7$$

$$3x = 33$$

$$\frac{\cancel{3}x}{\cancel{3}} = \frac{33}{3}$$

$$x = \$11$$

$$2x = \$22$$

Hence, the first server received $22, the second server received $11, and the third server received $7.

Step 4   Check: $22 + $11 + $7 = $40

PRACTICE

1. A stick 56 inches long is broken into two pieces so that one piece is 28 inches longer than the other piece. Find the length of each piece.
2. A person made two purchases. One purchase costs four times as much as the other purchase. The total cost of both purchases was $28. Find the cost of each purchase.
3. The difference between two numbers is 12. Find the values of each of the numbers if their sum is 28.
4. A mother is 4 times older than her daughter. In 14 years, she will be twice as old as her daughter. Find their ages now.
5. A house is worth four times as much as the lot it is built on. If the sum of the values of the house and lot is $156,000, find the value of each.

ANSWERS

1. 14 inches, 42 inches
2. $5.60, $22.40
3. 8, 20
4. mother's age = 28; daughter's age = 7
5. lot's value = $31,200; house's value = $124,800

Solving Percent Problems Using Equations

In Chapter 6 circles were used to solve percent problems. In Chapter 8 proportions will be used to solve percent problems. In this section, you will see that percent problems can be solved using equations.

To solve a percent problem, write an equation using the numbers and the word "of" as multiplication and the word "is" as an equal sign. The words "what number" or "what percent" translate to a variable such as x. Then solve the equation for x.

EXAMPLE
42% of 500 is what number?

SOLUTION
Translate the statement into an equation:

$$
\begin{array}{ccccc}
 & & & & \text{what} \\
42\% & \text{of} & 500 & \text{is} & \text{number?} \\
\downarrow & \downarrow & \downarrow & \downarrow & \downarrow \\
42\% & \cdot & 500 & = & x
\end{array}
$$

Solve the equation for x:

$$0.42 \times 500 = x$$
$$210 = x$$

(Be sure to change the percent to a decimal before multiplying.)

EXAMPLE

15 is what percent of 60?

SOLUTION

$$
\begin{array}{ccccc}
 & & \text{what} & & \\
15 & \text{is} & \text{percent} & \text{of} & 60? \\
\downarrow & \downarrow & \downarrow & \downarrow & \downarrow \\
15 & = & x & \cdot & 60
\end{array}
$$

$$\frac{15}{60} = \frac{x \cdot \cancel{60}}{\cancel{60}}$$
$$0.25 = x$$
$$25\% = x$$

EXAMPLE

90% of what number is 36?

SOLUTION

$$
\begin{array}{ccccc}
90\% & \text{of what number is } & 36? \\
\downarrow & \downarrow \;\; \downarrow \;\; \downarrow & \downarrow \\
90\% & \cdot \;\; x \;\; = & 36
\end{array}
$$

$$0.90x = 36$$
$$\frac{\cancel{0.90}x}{\cancel{0.90}} = \frac{36}{0.90}$$
$$x = 40$$

PRACTICE

Solve each percent problem using an equation:

1. 54% of 105 is what number?
2. 72% of what number is 360?
3. 55 is what percent of 440?
4. 6% of 88 is what number?
5. 35 is what percent of 350?

ANSWERS
1. 56.7
2. 500
3. 12.5%
4. 5.28
5. 10%

Quiz

1. Evaluate $-5xy$ when $x = -3$ and $y = 4$.
   (a) 12
   (b) 60
   (c) -12
   (d) -60

2. Evaluate $3(x^2 + 1)$ when $x = -5$.
   (a) 72
   (b) 108
   (c) 12
   (d) 78

3. Multiply $-4(2x - 6)$.
   (a) $-8x + 24$
   (b) $-8x - 24$
   (c) $-6x + 10$
   (d) $-8x - 6$

4. Multiply $2(x - y)$.
   (a) $2x - y$
   (b) $2x + y$
   (c) $2x - 2y$
   (d) $2x - y$

5. Combine like terms: $8x + 5x$.
   (a) $3x$
   (b) 13
   (c) $13x$
   (d) $13x^2$

6. Combine like terms: $-3x + 2 - 5y + 6x + 8$.
   (a) $-3x + 5y - 10$

(b) $11xy + 10$

(c) $2x - 3y + 5$

(d) $3x - 5y + 10$

7. Combine like terms: $5(x - 3) + 7x - 8$.
   (a) $2x - 5$
   (b) $12x - 23$
   (c) $12x - 11$
   (d) $-3x + 5$

8. Find the wing loading (L) in pounds per square foot of an airplane when the gross weight (W) is 5120 pounds and the wing area (A) is 320 square feet. Use $L = \frac{W}{A}$.
   (a) 27 pounds per square foot
   (b) 20 pounds per square foot
   (c) 16 pounds per square foot
   (d) 5 pounds per square foot

9. Find the distance (d) in feet an object falls when $g = -32$ and $t = 3$ seconds. Use $d = \frac{1}{2}gt^2$.
   (a) -96 feet
   (b) -144 feet
   (c) 48 feet
   (d) -288 feet

10. Solve $x + 52 = 38$.
    (a) -14
    (b) 90
    (c) 1976
    (d) 14

11. Solve $5x = 85$.
    (a) 15
    (b) 90
    (c) 17
    (d) 80

12. Solve $x - 16 = 84$.
    (a) 68
    (b) -68
    (c) 100
    (d) -100

13. Solve $\dfrac{x}{7} = 13$.
    (a) 2
    (b) 91
    (c) −78
    (d) 42

14. Solve $4x + 6 = -34$.
    (a) −10
    (b) 10
    (c) −7
    (d) 7

15. Solve $3(2x - 9) = 45$.
    (a) 9
    (b) −9
    (c) −12
    (d) 12

16. Solve $5x + 4 - 2x = 28 - 3x$
    (a) 8
    (b) 4
    (c) −4
    (d) −12

17. Represent algebraically two numbers so that one is 7 less than three times the other.
    (a) x, 3(x − 7)
    (b) x, 3x
    (c) x, 3x + 7
    (d) x, 3x −7

18. The sum of two numbers is 30. If one number is 8 more than the other number, find the numbers.
    (a) 8, 22
    (b) 7, 23
    (c) 11, 19
    (d) 8, 16

19. What percent of 32 is 24? Solve using an equation.
    (a) 25%
    (b) 133.$\overline{3}$%
    (c) 125%
    (d) 75%

20. 60% of what number is 42? Solve using an equation.
    (a) 70
    (b) 25.2
    (c) 105
    (d) 7

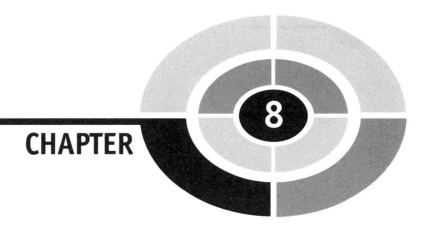

Ratio and Proportion

Ratio

A **ratio** is a comparison of two numbers. For example, consider the statement, "About 14 out of every 100 people in the United States do not have health insurance."

A ratio can be expressed by a fraction or by using a colon. In the preceding example, the ratio $\frac{14}{100}$ is the same as 14 to 100 or 14 : 100. Fractions are usually reduced to lowest terms, so the ratio becomes $\frac{7}{50}$ or 7 : 50.

It is important to understand that whatever number comes first in the ratio statement is placed in the numerator of the fraction and whatever number comes second in the ratio statement is placed in the denominator of the fraction. In general, the ratio of a to b is written as $\frac{a}{b}$.

EXAMPLE
Find the ratio of 3 to 12.

SOLUTION

$$3 \text{ to } 12 = \frac{3}{12} = \frac{1}{4}$$

EXAMPLE

If 8 oranges cost $2.00, find the ratio of oranges to cost.

SOLUTION

$$\frac{8 \text{ oranges}}{2 \text{ dollars}} = \frac{4 \text{ oranges}}{1 \text{ dollar}} = \frac{4 \text{ oranges}}{\$1.00}$$

PRACTICE

1. Find the ratio of 10 to 25.
2. Find the ratio of 9 to 5.
3. Find the ratio of 80 miles to 2 hours.
4. Find the ratio of 2 yards to 60 cents.
5. Find the ratio of one nickel to one quarter.

ANSWERS

1. $\dfrac{2}{5}$

2. $\dfrac{9}{5}$

3. $\dfrac{40 \text{ miles}}{1 \text{ hour}}$

4. $\dfrac{1 \text{ yard}}{30 \text{ cents}}$

5. $\dfrac{1}{5}$

Proportion

A **proportion** is a statement of equality of two ratios. For example, $\frac{3}{4} = \frac{6}{8}$ is a proportion. Proportions can also be expressed using a colon, as $3:4 = 6:8$.

A proportion consists of 4 terms. Usually it is necessary to find the value of one term given the other three terms. For example, $\frac{5}{8} = \frac{x}{24}$. In order to do this, it is necessary to use a basic rule of proportions:

$$\frac{a}{b} = \frac{c}{d} \quad \text{if} \quad a \cdot d = b \cdot c$$

In other words, $\frac{3}{4} = \frac{6}{8}$ if $3 \times 8 = 4 \times 6$ or $24 = 24$. This is called **cross multiplication**.

In order to find the unknown in a proportion, cross multiply and then solve the equation for the unknown: i.e., divide by the number that is in front of the variable.

EXAMPLE
Find the value of x:

$$\frac{x}{5} = \frac{36}{45}$$

SOLUTION

$$\frac{x}{5} = \frac{36}{45}$$

$$45x = 5 \cdot 36 \qquad \text{Cross multiply}$$

$$45x = 180$$

$$\frac{\cancel{45}x}{\cancel{45}} = \frac{180}{45} \qquad \text{Divide}$$

$$x = 4$$

EXAMPLE
Find the value of x:

$$\frac{24}{32} = \frac{x}{4}$$

SOLUTION

$$\frac{24}{32} = \frac{x}{4}$$

$$24 \cdot 4 = 32x \qquad \text{Cross multiply}$$

$$96 = 32x$$

$$\frac{96}{32} = \frac{\cancel{32}x}{\cancel{32}} \qquad \text{Divide}$$

$$3 = x$$

EXAMPLE
Find the value of x:

$$\frac{x}{15} = \frac{26}{75}$$

SOLUTION

$$\frac{x}{15} = \frac{26}{75}$$

$75x = 15 \cdot 26$ Cross multiply

$75x = 390$

$$\frac{\cancel{75}x}{\cancel{75}} = \frac{390}{75}$$ Divide by 75

$$x = 5\tfrac{1}{5} \text{ or } 5.2$$

PRACTICE

Find the value of x:

1. $\dfrac{3}{4} = \dfrac{24}{x}$

2. $\dfrac{5}{x} = \dfrac{35}{42}$

3. $\dfrac{12}{7} = \dfrac{x}{56}$

4. $\dfrac{25}{8} = \dfrac{x}{56}$

5. $\dfrac{x}{9} = \dfrac{22.75}{63}$

ANSWERS

1. 32
2. 6
3. 96
4. 175
5. 3.25

Word Problems

In order to solve word problems using proportions:

- *Step 1   Read the problem.*
- *Step 2   Identify the ratio statement.*
- *Step 3   Set up the proportion.*
- *Step 4   Solve for x.*

In problems involving proportions, there will always be a ratio statement. It is important to find the ratio statement and then set up the proportion using the ratio statement. Be sure to keep the identical units in the numerators and denominators of the fractions in the proportion.

EXAMPLE
An automobile travels 176 miles on 8 gallons of gasoline. How far can it go on a tankful of gasoline if the tank holds 14 gallons?

SOLUTION
The ratio statement is:

$$\frac{176 \text{ miles}}{8 \text{ gallons}}$$

so the proportion is:

$$\frac{176 \text{ miles}}{8 \text{ gallons}} = \frac{x \text{ miles}}{14 \text{ gallons}}$$

Solve:

$$\frac{176}{8} = \frac{x}{14}$$

$$8x = 14 \cdot 176$$

$$8x = 2464$$

$$\frac{8x}{8} = \frac{2464}{8}$$

$$x = 308 \text{ miles}$$

Hence, on 14 gallons, the automobile can travel a distance of 308 miles.

Math Note: Notice in the previous example that the numerators of the proportions have the same units, miles, and the denominators have the same units, gallons.

EXAMPLE
If it takes 16 yards of material to make 3 costumes of a certain size, how much material will be needed to make 8 costumes of that same size?

SOLUTION

The ratio statement is:

$$\frac{16 \text{ yards}}{3 \text{ costumes}}$$

The proportion is:

$$\frac{16 \text{ yards}}{3 \text{ costumes}} = \frac{x \text{ yards}}{8 \text{ costumes}}$$

$$\frac{16}{3} = \frac{x}{8}$$

$$16 \cdot 8 = 3x$$

$$128 = 3x$$

$$\frac{128}{3} = \frac{\cancel{3}x}{\cancel{3}}$$

$$42\tfrac{2}{3} \text{ yards} = x$$

Hence, $42\tfrac{2}{3}$ yards or $42.\overline{6}$ yards should be purchased.

PRACTICE

1. If 5 pounds of grass seed will cover 1025 square feet, how many square feet can be covered by 15 pounds of grass seed?
2. If a homeowner pays $8000 a year in taxes for a house valued at $250,000, how much would a homeowner pay in yearly taxes on a house valued at $175,000?
3. If a person earns $2340 in 6 weeks, how much can the person earn in 13 weeks at the same rate of pay?
4. If 3 gallons of waterproofing solution can cover 360 square feet of decking, how much solution will be needed to cover a deck that is 2520 square feet in size?
5. A recipe for 4 servings requires 6 tablespoons of shortening. If the chef wants to make enough for 9 servings, how many tablespoons of shortening are needed?

ANSWERS

1. 3075 square feet
2. $5600
3. $5070
4. 21 gallons
5. $13\tfrac{1}{2}$ or 13.5 tablespoons

Solving Percent Problems Using Proportions

Recall that percent means part of 100, so a percent can be written as a ratio using 100 as the denominator of the fraction. For example, 43% can be written as $\frac{43}{100}$. Also, recall that a percent problem has a base, B, a part, P, and a ratio (percent) R, so a proportion can be set up for each type of percent problem as:

$$\frac{R\%}{100\%} = \frac{P}{B}$$

EXAMPLE
Find 32% of 48.

SOLUTION

$$\frac{32\%}{100\%} = \frac{P}{48}$$
$$32 \cdot 48 = 100P$$
$$1536 = 100P$$
$$\frac{1536}{100} = \frac{\cancel{100}P}{\cancel{100}}$$
$$15.36 = P$$

EXAMPLE
What percent of 60 is 45?

SOLUTION

$$\frac{R}{100\%} = \frac{45}{60}$$
$$60R = 45 \cdot 100$$
$$60R = 4500$$
$$\frac{\cancel{60}R}{\cancel{60}} = \frac{4500}{60}$$
$$R = 75, \text{ so the rate is } 75\%.$$

EXAMPLE
22% of what number is 17.6?

SOLUTION

$$\frac{22\%}{100\%} = \frac{17.6}{B}$$
$$22B = 17.6 \cdot 100$$
$$22B = 1760$$
$$\frac{\cancel{22}B}{\cancel{22}} = \frac{1760}{22}$$
$$B = 80$$

PRACTICE
Solve each percent problem by using a proportion.

1. Find 54% of 80.
2. 6 is what percent of 25?
3. 25% of what number is 64?
4. 24 is what percent of 120?
5. Find 92% of 450.
6. 42% of what number is 35.28?
7. 15 is what percent of 60?
8. Find 36% of 72.
9. 18% of what number is 10.08?
10. 12 is what percent of 30?

ANSWERS
1. 43.2
2. 24%
3. 256
4. 20%
5. 414
6. 84
7. 25%
8. 25.92
9. 56
10. 40%

Quiz

1. Find the ratio of 2 to 7.

(a) $\dfrac{7}{2}$

(b) $\dfrac{2}{9}$

(c) $\dfrac{2}{7}$

(d) $\dfrac{7}{9}$

2. Find the ratio of 6 to 18.
   (a) $1:3$
   (b) $6:24$
   (c) $18:24$
   (d) $3:1$

3. Find the ratio of 12 to 3.

   (a) $\dfrac{1}{4}$

   (b) $\dfrac{3}{12}$

   (c) $\dfrac{4}{1}$

   (d) $\dfrac{12}{15}$

4. Find the ratio of 5 to 9.
   (a) $5:14$
   (b) $9:14$
   (c) $9:5$
   (d) $5:9$

5. Find the ratio of 100 to 40.

   (a) $\dfrac{40}{100}$

   (b) $\dfrac{100}{140}$

   (c) $\dfrac{2}{5}$

(d) $\dfrac{5}{2}$

6. Find the value of x: $\dfrac{8}{9} = \dfrac{x}{27}$.

   (a) 16
   (b) 12
   (c) 8
   (d) 24

7. Find the value of x: $\dfrac{1}{3} = \dfrac{13}{x}$.

   (a) 39
   (b) 3
   (c) 13
   (d) 26

8. Find the value of x: $\dfrac{x}{6} = \dfrac{8}{12}$.

   (a) 3
   (b) 4
   (c) 6
   (d) 12

9. Find the value of x: $\dfrac{x}{5} = \dfrac{5}{8}$.

   (a) 8
   (b) 3.125
   (c) 5
   (d) 6.375

10. Find the value of x: $\dfrac{0.6}{0.15} = \dfrac{x}{0.3}$.

    (a) 1.6
    (b) 1.5
    (c) 1.2
    (d) 1.8

11. Find the value of x: $\dfrac{x}{0.8} = \dfrac{0.5}{0.2}$.

    (a) 20
    (b) 0.2
    (c) 2
    (d) 0.02

12. Find the value of x: $\dfrac{5}{x} = \dfrac{6}{10}$.

    (a) $8\frac{2}{3}$

(b) $8\frac{1}{3}$

(c) 3

(d) 12

13. On a scale drawing, 3 inches represents 24 feet. How tall is a building that is 5 inches high?
(a) 20 feet
(b) 30 feet
(c) 40 feet
(d) 50 feet

14. If 7 pounds of candy costs \$8.40, how much will 18 pounds cost?
(a) \$21.60
(b) \$2.57
(c) \$151.20
(d) \$24.70

15. A concrete mixture calls for one part cement to three parts sand. How much cement is needed if the contractor used 22 buckets of sand?

(a) $5\frac{2}{3}$ buckets

(b) $6\frac{3}{4}$ buckets

(c) $7\frac{1}{3}$ buckets

(d) 33 buckets

16. If a picture 4 inches wide by 6 inches long is to be enlarged so that the length is 24 inches, what would be the size of the width?
(a) 8 inches
(b) 10 inches
(c) 12 inches
(d) 16 inches

17. If 25 feet of a certain type of cable weighs $3\frac{1}{8}$ pounds, how much will 56 feet of cable weigh?

(a) $6\frac{1}{4}$ pounds

(b) 7 pounds

(c) $8\frac{1}{2}$ pounds

(d) 9 pounds

18. What percent of 16 is 10?
    (a) 62.5%
    (b) 37.5%
    (c) 12.5%
    (d) 78.5%

19. Solve using a proportion: 32% of 50 is what number?
    (a) 8
    (b) 16
    (c) 24
    (d) 30

20. Solve using a proportion: 16% of what number is 48?
    (a) 3
    (b) 30
    (c) 300
    (d) 3000

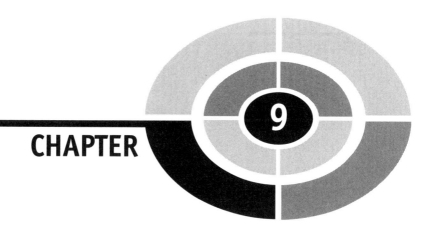

CHAPTER

Informal Geometry

Geometric Figures

The word "geometry" is derived from two Greek words meaning "earth measure." The basic geometric figures are the point, the line, and the plane. These figures are theoretical and cannot be formally defined. A **point** is represented by a dot and is named by a capital letter. A **line** is an infinite set of points and is named by a small letter or by two points on the line. A **line segment** is part of a line between two points called endpoints. A **plane** is a flat surface (see Fig. 9-1).

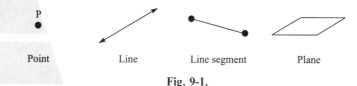

| Point | Line | Line segment | Plane |

Fig. 9-1.

Points and line segments are used to make geometric figures. The geometric figures presented in this chapter are the triangle, the square, the rectangle, the parallelogram, the trapezoid, and the circle (see Fig. 9-2).

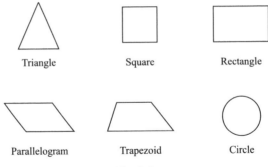

Fig. 9-2.

A **triangle** is a geometric figure with three sides. A **rectangle** is a geometric figure with four sides and four 90° angles. The opposite sides are equal in length and are parallel. A **square** is a rectangle in which all sides are the same length. A **parallelogram** has four sides with two pairs of parallel sides. A **trapezoid** has four sides, two of which are parallel. A **circle** is a geometric figure such that all the points are the same distance from a point called its center. The center is not part of the circle. A line segment passing through the center of a circle and with its endpoints on the circle is called a **diameter**. A line segment from the center of a circle to the circle is called a **radius** (see Fig. 9-3).

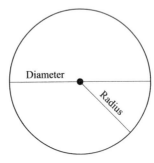

Fig. 9-3.

Perimeter

The distance around the outside of a geometric figure is called the **perimeter** of the figure. The perimeter is found by adding the measures of all sides of the geometric figure. For some geometric figures, there are special formulas that are used to find their perimeters.

The perimeter of a triangle is found by adding the lengths of the three sides:
P = a + b + c.

EXAMPLE
Find the perimeter of the triangle shown in Fig. 9-4.

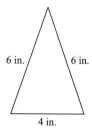

6 in.   6 in.

4 in.

Fig. 9-4.

SOLUTION

$$P = a + b + c$$
$$= 6 \text{ inches } + 6 \text{ inches } + 4 \text{ inches}$$
$$= 16 \text{ inches}$$

The perimeter of a rectangle can be found by using the formula P = 2l + 2w,
where l is the length and w is the width.

EXAMPLE
Find the perimeter of the rectangle shown in Fig. 9-5.

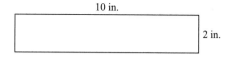

10 in.

2 in.

Fig. 9-5.

SOLUTION

$$P = 2l + 2w$$
$$= 2 \cdot 10 \text{ inches} + 2 \cdot 2 \text{ inches}$$
$$= 20 + 4$$
$$= 24 \text{ inches}$$

The perimeter of a square is found by using the formula P = 4s, where s is the length of a side.

EXAMPLE

Find the perimeter of the square shown in Fig. 9-6.

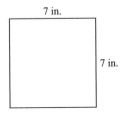

7 in.

7 in.

Fig. 9-6.

SOLUTION

$$P = 4s$$
$$= 4 \cdot 7 \text{ inches}$$
$$= 28 \text{ inches}$$

The perimeter of a circle is called the **circumference** of a circle. To find the circumference of a circle, it is necessary to use either the radius (r) or the diameter (d) of the circle.

The radius of a circle is equal to one-half of its diameter: i.e., $r = \frac{1}{2}d$ or $\frac{d}{2}$. The diameter of a circle is twice as large as its radius: i.e., $d = 2r$.

EXAMPLE

Find the radius of a circle if its diameter is 16 inches.

SOLUTION

$$r = \frac{1}{2}d$$
$$= \frac{1}{2} \cdot 16 \text{ inches}$$
$$= 8 \text{ inches}$$

EXAMPLE

Find the diameter of a circle when its radius is 14 inches.

SOLUTION

$$d = 2r$$
$$= 2 \cdot 14 \text{ inches}$$
$$= 28 \text{ inches}$$

In order to find the circumference of a circle, a special number is used. That number is called pi, and its symbol is π, where $\pi = 3.141592654\ldots$. For our purposes, we use π rounded to $3.14 : \pi$ is the number of times that a diameter of a circle will fit around the circle.

The circumference of a circle then is $C = \pi d$ or $C = 2\pi r$, where r is the radius of the circle.

EXAMPLE
Find the circumference of a circle whose diameter is 7 inches. Use $\pi = 3.14$.

SOLUTION

$$C = \pi d$$
$$= 3.14 \cdot 7 \text{ inches}$$
$$= 21.98 \text{ inches}$$

EXAMPLE
Find the circumference of a circle whose radius is 12 feet. Use $\pi = 3.14$.

SOLUTION

$$C = 2\pi r$$
$$= 2 \cdot 3.14 \cdot 12 \text{ feet}$$
$$= 75.36 \text{ feet}$$

Math Note: Sometimes $\frac{22}{7}$ is used as an approximate value for π.

PRACTICE
1. Find the perimeter of a triangle whose sides are 8 feet, 9 feet, and 11 feet.
2. Find the perimeter of a square whose side is 3 yards.
3. Find the perimeter of a rectangle whose length is 28 inches and whose width is 16 inches.

4. Find the circumference of a circle whose diameter is 12 inches. Use $\pi = 3.14$.

5. Find the circumference of a circle whose radius is 30 feet. Use $\pi = 3.14$.

ANSWERS

1. 28 feet
2. 12 yards
3. 88 inches
4. 37.68 inches
5. 188.4 feet

Area

The **area** of a geometric figure is the number of square units contained in its surface. For example, the area of the 3-inch by 2-inch rectangle is 6 square inches as shown in Fig. 9-7.

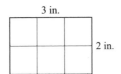

3 in.

2 in.

Fig. 9-7.

Area is measured in square units. A square inch is a square whose sides measure one inch. A square foot is a square whose sides measure one foot, etc. Square units are abbreviated using two for the exponent:

- 1 square inch = 1 in.^{2}
- 1 square foot = 1 ft^2
- 1 square yard = 1 yd^2
- 1 square mile = 1 mi^2

The area of a rectangle is found by using the formula $A = lw$, where l is the length and w is the width.

EXAMPLE

Find the area of the rectangle shown in Fig. 9-8.

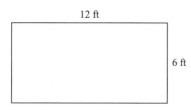

Fig. 9-8.

SOLUTION

$$A = lw$$
$$= 12 \text{ ft} \times 6 \text{ ft}$$
$$= 72 \text{ ft}^2$$

The area of a square can be found by using the formula $A = s^2$, where s is the length of its side.

EXAMPLE
Find the area of the square shown in Fig. 9-9.

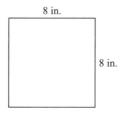

Fig. 9-9.

SOLUTION

$$A = s^2$$
$$= (8 \text{ in.})^2$$
$$= 64 \text{ in.}^2$$

To find the area of a triangle you need to know the measure of its *altitude*. The **altitude** of a triangle is the measure of a perpendicular line from its highest point to its base (the side opposite its highest point). See Fig. 9-10.

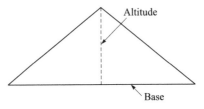

Fig. 9-10.

The area of a triangle can be found by using the formula $A = \frac{1}{2}bh$, where b is the base and h is the height or altitude.

EXAMPLE
Find the area of the triangle shown in Fig. 9-11.

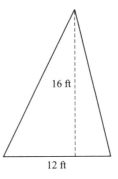

16 ft

12 ft

Fig. 9-11.

SOLUTION

$$A = \frac{1}{2}bh$$

$$= \frac{1}{2} \cdot 12 \text{ ft} \cdot 16 \text{ ft}$$

$$= 96 \text{ ft}^2$$

The area of a parallelogram can be found by using the formula $A = bh$, where b is the base and h is the height or altitude.

EXAMPLE
Find the area of the parallelogram shown in Fig. 9-12.

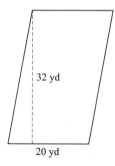

Fig. 9-12.

SOLUTION

$$A = bh$$
$$= 20 \text{ yd} \cdot 32 \text{ yd}$$
$$= 640 \text{ yd}^2$$

A trapezoid has a height and two bases — a lower base, b_1, and an upper base, b_2 (see Fig. 9-13).

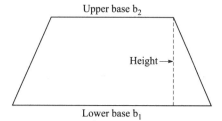

Fig. 9-13.

The area of a trapezoid can be found by using the formula $A = \frac{1}{2} h(b_1 + b_2)$

EXAMPLE

Find the area of the trapezoid shown in Fig. 9-14.

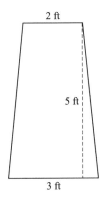

2 ft

5 ft

3 ft

Fig. 9-14.

SOLUTION

$$A = \frac{1}{2}h(b_1 + b_2)$$

$$= \frac{1}{2} \cdot 5 \text{ ft} \cdot (3 \text{ ft} + 2 \text{ ft})$$

$$= \frac{1}{2} \cdot 5 \text{ ft} \cdot 5 \text{ ft}$$

$$= 12.5 \text{ ft}^2$$

The area of a circle can be found by using the formula $A = \pi r^2$, where $\pi = 3.14$ and r is the radius.

EXAMPLE

Find the area of the circle shown in Fig. 9-15. Use $\pi = 3.14$.

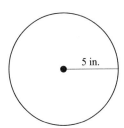

5 in.

Fig. 9-15.

SOLUTION

$$A = \pi r^2$$
$$= 3.14 \cdot (5 \text{ in.})^2$$
$$= 3.14 \cdot 25 \text{ in.}^2$$
$$= 78.5 \text{ in.}^2$$

The following suggestions will help you to convert between units. To change:

- square feet to square inches, multiply by 144;
- square inches to square feet, divide by 144;
- square yards to square feet, multiply by 9;
- square feet to square yards, divide by 9.

EXAMPLE
Change 1728 square inches to square feet.

SOLUTION

$$1728 \text{ in.}^2 \div 144 = 12 \text{ ft}^2$$

EXAMPLE
Change 15 square yards to square feet.

SOLUTION

$$15 \text{ yd}^2 \times 9 = 135 \text{ ft}^2$$

PRACTICE
1. Find the area of a rectangle whose length is 123 yards and whose width is 53 yards.
2. Find the area of a square whose side is 53 inches.
3. Find the area of a triangle whose base is 16 inches and whose height is 9 inches.
4. Find the area of a parallelogram whose base is 22 inches and whose height is 10 inches.
5. Find the area of a trapezoid whose bases are 32 feet and 44 feet and whose height is 20 feet.
6. Find the area of a circle whose radius is 14 inches. Use $\pi = 3.14$.
7. Find the area of a circle whose diameter is 32 feet. Use $\pi = 3.14$.
8. Change 12 square feet to square inches.

9. Change 27 square feet to square yards.
10. Change 2 square yards to square inches.

ANSWERS
  1. 6519 yd^2
  2. 2809 in.^{2}
  3. 72 in.^{2}
  4. 220 in.^{2}
  5. 760 ft^2
  6. 615.44 in.^{2}
  7. 803.84 ft^2
  8. 1728 in.^{2}
  9. 3 yd^2
 10. 2592 in.^{2}

Volume

The volume of a geometric figure is a measure of its capacity. Volume is measured in cubic units. Cubic units are abbreviated using 3 for the exponent:

- 1 cubic inch = 1 in.^{3}
- 1 cubic foot = 1 ft^3
- 1 cubic yard = 1 yd^3

The basic geometric solids are the rectangular solid, the cube, the cylinder, the sphere, the right circular cone, and the pyramid. These figures are shown in Fig. 9-16.

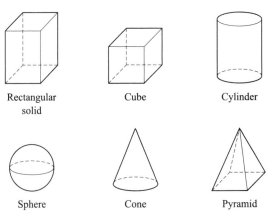

Rectangular   Cube   Cylinder
solid

Sphere   Cone   Pyramid

Fig. 9-16.

The volume of a rectangular solid can be found by using the formula V = lwh, where l = the length, w = the width, and h = the height.

EXAMPLE

Find the volume of the rectangular solid shown in Fig. 9-17.

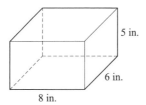

5 in.

6 in.

8 in.

Fig. 9-17.

SOLUTION

$$V = lwh$$
$$= 8 \text{ in.} \cdot 6 \text{ in.} \cdot 5 \text{ in.}$$
$$= 240 \text{ in.}^3$$

The volume of a cube can be found by using the formula V = s^3, where s is the length of the side.

EXAMPLE

Find the volume of the cube shown in Fig. 9-18.

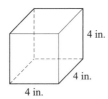

4 in.

4 in.

4 in.

Fig. 9-18.

SOLUTION

$$V = s^3$$
$$= (4 \text{ in.})^3$$
$$= 64 \text{ in.}^3$$

The volume of a cylinder can be found by using the formula $V = \pi r^2 h$, where r is the radius of the base and h is the height.

EXAMPLE
Find the volume of the cylinder shown in Figure 9-19. Use $\pi = 3.14$.

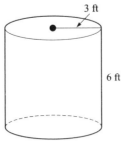

3 ft

6 ft

Fig. 9-19.

SOLUTION

$$V = \pi r^2 h$$
$$= 3.14 \cdot (3 \text{ ft})^2 \cdot 6 \text{ ft}$$
$$= 169.56 \text{ ft}^3$$

The volume of a sphere can be found by using the formula $V = \frac{4}{3}\pi r^3$, where r is the radius of the sphere.

EXAMPLE
Find the volume of the sphere shown in Figure 9-20. Use $\pi = 3.14$.

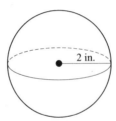

2 in.

Fig. 9-20.

SOLUTION

$$V = \frac{4}{3}\pi r^3$$

$$= \frac{4}{3} \cdot 3.14 \cdot (2 \text{ in.})^3$$

$$= 33.49 \text{ in.}^3 \text{ (rounded)}$$

The volume of a right circular cone can be found by using the formula $V = \frac{1}{3}\pi r^2 h$, where r is the radius of the base and h is the height of the cone.

EXAMPLE

Find the volume of the cone shown in Fig. 9-21. Use $\pi = 3.14$.

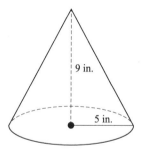

9 in.

5 in.

Fig. 9-21.

SOLUTION

$$V = \frac{1}{3}\pi r^2 h$$

$$= \frac{1}{3} \cdot 3.14 \cdot (5 \text{ in.})^2 \cdot 9 \text{ in.}$$

$$= 235.5 \text{ in.}^3$$

The volume of a pyramid can be found by using the formula $V = \frac{1}{3}Bh$, where B is the area of the base and h is the height of the pyramid. If the base is a square, use $B = s^2$. If the base is a rectangle, use $B = lw$.

EXAMPLE

Find the volume of the pyramid shown in Fig. 9-22.

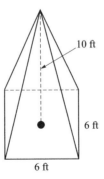

Fig. 9-22.

SOLUTION

In this case, the base is a square, so the area of the base is $B = s^2$.

$$V = \frac{1}{3} B \cdot h$$

$$= \frac{1}{3} (6 \text{ ft})^2 \cdot 10 \text{ ft}$$

$$= 120 \text{ ft}^3$$

Sometimes it is necessary to convert from cubic yards to cubic feet, cubic inches to cubic feet, etc. The following information will help you to do this. To change:

- cubic feet to cubic inches, multiply by 1728;
- cubic inches to cubic feet, divide by 1728;
- cubic yards to cubic feet, multiply by 27;
- cubic feet to cubic yards, divide by 27.

EXAMPLE

Change 18 cubic yards to cubic feet.

SOLUTION

$$18 \times 27 = 486 \text{ cubic feet}$$

EXAMPLE

Change 15,552 cubic inches to cubic feet.

SOLUTION

$$15,552 \div 1728 = 9 \text{ cubic feet}$$

PRACTICE
1. Find the volume of a rectangular solid if it is 8 inches long, 6 inches wide, and 10 inches high.
2. Find the volume of a cube if the side is 13 feet.
3. Find the volume of a cylinder if the radius is 3 inches and its height is 7 inches. Use $\pi = 3.14$.
4. Find the volume of a sphere if its radius is 9 inches. Use $\pi = 3.14$.
5. Find the volume of a cone if its radius is 6 feet and its height is 8 feet. Use $\pi = 3.14$.
6. Find the volume of a pyramid if its height is 2 yards and its base is a rectangle whose length is 3 yards and whose width is 2.5 yards.

ANSWERS
1. 480 in.^{3}
2. 2197 ft^3
3. 197.82 in.^{3}
4. 3052.08 in.^{3}
5. 301.44 ft^3
6. 5 yd^3

Word Problems

Word problems in geometry can be solved by the following procedure:

- *Step 1   Read the problem.*
- *Step 2   Decide whether you are being asked to find the perimeter, area, or volume.*
- *Step 3   Select the correct formula.*
- *Step 4   Substitute in the formula and evaluate.*

EXAMPLE
How many inches of fringe are needed for a quilt that measures 110 inches by 72 inches?

SOLUTION
Since the fringe goes along the edge, it is necessary to find the perimeter of a rectangle 110 inches by 72 inches:

$$P = 2l + 2w$$
$$= 2 \cdot 110 \text{ in.} + 2 \cdot 72 \text{ in.}$$
$$= 364 \text{ in.}$$

EXAMPLE

How many cubic yards of dirt must be removed from a rectangular foundation for a house that measures 54 feet by 28 feet by 8 feet?

SOLUTION

Find the volume of the rectangular solid:

$$V = lwh$$
$$= 54 \text{ ft} \cdot 28 \text{ ft} \cdot 8 \text{ ft}$$
$$= 12,096 \text{ ft}^3$$

Since the problem asks for cubic yards, change 12,096 cubic feet to cubic yards:

$$12,096 \div 27 = 448 \text{ yd}^3$$

PRACTICE

1. A silo is 32 feet high and has a diameter of 10 feet. How many cubic feet will it hold? Use $\pi = 3.14$.
2. If a rug sells for $35 a square yard, how much will it cost to cover the floor of a room that measures 12 feet by 15 feet?
3. If a revolving lawn sprinkler will spray a distance of 12 feet, how many square feet will the sprinkler cover in one revolution? Use $\pi = 3.14$.
4. How many cubic yards of air would be in a balloon that has a 30 foot diameter? Use $\pi = 3.14$.
5. How many decorative bricks 10 inches long would be needed to go around a circular garden that is 8 feet in diameter? Use $\pi = 3.14$.

ANSWERS

1. 2512 ft^3
2. $700
3. 452.16 ft^2
4. 523.33 yd^3 (rounded)
5. About 30

Square Roots

In Chapter 2 you learned how to square a number. For example, $7^2 = 7 \times 7 = 49$ and $3^2 = 3 \times 3 = 9$. The opposite of squaring a number is taking the square root of a number. The radical sign ($\sqrt{}$) is used to indicate the square root of a number, so $\sqrt{16} = 4$ because $4^2 = 16$ and $\sqrt{25} = 5$ because $5^2 = 25$.

Numbers such as 1, 4, 9, 16, 25, 36, 49, etc., are called perfect squares because their square roots are rational numbers.

The square roots of other numbers such as $\sqrt{2}$, $\sqrt{3}$, $\sqrt{5}$, etc., are called **irrational** numbers because their square roots are infinite, non-repeating decimals. For example $\sqrt{2} = 1.414213562 \ldots$ and $\sqrt{3} = 1.732050808 \ldots$. In other words, the decimal value of $\sqrt{2}$ cannot be found exactly.

Math Note: The easiest way to find the square root of a number is to use a calculator. However, the square root of a perfect square can be found by guessing and then squaring the answer to see if it is correct. For example, to find $\sqrt{196}$, you could guess it is 12. Then square 12. $12^2 = 144$. This is too small. Try 13. $13^2 = 169$. This is still too small, so try 14. $14^2 = 196$. Hence, $\sqrt{196} = 14$.

The set of numbers which consists of the rational numbers and the irrational numbers is called the **real numbers.**

The Pythagorean Theorem

The Pythagorean theorem, an important mathematical principle, uses right triangles. A **right triangle** is a triangle which has one right or 90° angle. The side opposite the 90° angle is called the **hypotenuse**.

The Pythagorean theorem states that for any right triangle $c^2 = a^2 + b^2$, where c is the length of the hypotenuse and a and b are the lengths of its sides (see Fig. 9-23).

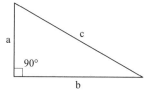

Fig. 9-23.

If you need to find the hypotenuse of a right triangle, use $c = \sqrt{a^2 + b^2}$. If you need to find the length of one side of a right triangle, use $a = \sqrt{c^2 - b^2}$ or $b = \sqrt{c^2 - a^2}$.

EXAMPLE
Find the length of the hypotenuse of a right triangle if the length of its sides are 5 inches and 12 inches.

SOLUTION

$$c = \sqrt{a^2 + b^2}$$
$$= \sqrt{5^2 + 12^2}$$
$$= \sqrt{25 + 144}$$
$$= \sqrt{169}$$
$$= 13 \text{ inches}$$

EXAMPLE
A person drives 33 miles due east, then makes a left turn, and drives 56 miles north. How far is he from his starting point?

SOLUTION
The right triangle is shown in Fig. 9-24.

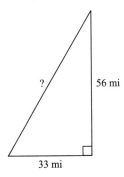

Fig. 9-24.

Then

$$c = \sqrt{a^2 + b^2}$$
$$= \sqrt{33^2 + 56^2}$$
$$= \sqrt{1089 + 3136}$$
$$= \sqrt{4225}$$
$$= 65 \text{ miles}$$

PRACTICE

1. Find the length of the hypotenuse of a right triangle whose sides are 24 inches and 18 inches.
2. Find the length of a side of a right triangle whose hypotenuse is 17 yards and whose other side is 15 yards.
3. Find the distance that you would walk up a staircase if it is 21 feet high and has a base of 28 feet.
4. Find the perimeter of a flower garden if its shape is a right triangle with a hypotenuse of 10 feet and a side of 8 feet.
5. Find the hypotenuse of a sail whose shape is a right triangle with a base of 15 feet and whose height is 36 feet.

ANSWERS

1. $\sqrt{900} = 30$ inches
2. $\sqrt{64} = 8$ yards
3. $\sqrt{1225} = 35$ feet
4. 24 feet
5. $\sqrt{1521} = 39$ feet

Quiz

1. Find the perimeter of a rectangle whose length is 18 feet and whose width is 7 feet.
   (a) 11 ft
   (b) 25 ft
   (c) 50 ft
   (d) 126 ft

2. Find the circumference of a circle whose diameter is 24 inches. Use $\pi = 3.14$.
   (a) 37.68 in.
   (b) 75.36 in.
   (c) 452.16 in.
   (d) 1808.64 in.

3. Find the perimeter of a square whose side is 19 inches.
   (a) 361 in.
   (b) 57 in.

(c) 38 in.
(d) 76 in.

4. Find the perimeter of a triangle whose sides are 12 yards, 13 yards, and 16 yards.
   (a) 41 yd
   (b) 82 yd
   (c) 20.5 yd
   (d) 123 yd

5. If the radius of a circle is 4.8 inches, find the diameter.
   (a) 9.6 in.
   (b) 15.072 in.
   (c) 4.8 in.
   (d) 2.4 in.

6. Find the area of a circle whose radius is 9 inches. Use $\pi = 3.14$.
   (a) 28.26 in.^{2}
   (b) 254.34 in.^{2}
   (c) 1017.36 in.^{2}
   (d) 56.52 in.^{2}

7. Find the area of a square whose side is 6.2 feet.
   (a) 38.44 ft^2
   (b) 24.8 ft^2
   (c) 1.55 ft^2
   (d) 12.4 ft^2

8. Find the area of a trapezoid whose height is 12 inches and whose bases are 7 inches and 14 inches.
   (a) 288 in.^{2}
   (b) 252 in.^{2}
   (c) 126 in.^{2}
   (d) 144 in.^{2}

9. Find the area of a rectangle whose length is 8 inches and whose width is 5 inches.
   (a) 40 in.^{2}
   (b) 26 in.^{2}
   (c) 13 in^2
   (d) 25 in.^{2}

10. Find the area of a triangle whose base is 42 yards and whose height is 36 yards.

(a) 78 yd^2
(b) 39 yd^2
(c) 756 yd^2
(d) 1512 yd^2

11. Find the area of a parallelogram whose base is 10 inches and whose height is 5 inches.
(a) 30 in.^{2}
(b) 50 in.^{2}
(c) 15 in.^{2}
(d) 25 in.^{2}

12. How many square inches are there in 15 square feet?
(a) 2160 in.^{2}
(b) 180 in.^{2}
(c) 225 in.^{2}
(d) 135 in.^{2}

13. How many square yards are there in 4860 square feet?
(a) 1620 yd^2
(b) 540 yd^2
(c) 810 yd^2
(d) 405 yd^2

14. Find the volume of a sphere whose radius is 9 inches. Use $\pi = 3.14$.
(a) 28.26 in.^{3}
(b) 3052.08 in.^{3}
(c) 972 in.^{3}
(d) 2289.06 in.^{3}

15. Find the volume of a cube whose side is 3 yards in length.
(a) 9 yd^3
(b) 12 yd^3
(c) 18 yd^3
(d) 27 yd^3

16. Find the volume of a cylinder whose base has a radius of 4 inches and whose height is 9 inches. Use $\pi = 3.14$.
(a) 452.16 in.^{3}
(b) 226.08 in.^{3}
(c) 113.04 in.^{3}
(d) 1017.36 in.^{3}

17. Find the volume of a rectangular solid whose length is 14 inches, whose width is 5 inches, and whose length is 7 inches.
    (a) 26 in.^{3}
    (b) 52 in.^{3}
    (c) 490 in.^{3}
    (d) 980 in.^{3}

18. Find the volume of a cone whose base has a diameter of 8 feet and whose height is 12 feet. Use $\pi = 3.14$.
    (a) 100.48 ft^3
    (b) 200.96 ft^3
    (c) 803.84 ft^3
    (d) 1205.76 ft^3

19. Find the volume of a pyramid whose height is 3 feet and whose base is a square with a side of 4 feet.
    (a) 4 ft^3
    (b) 8 ft^3
    (c) 12 ft^3
    (d) 16 ft^3

20. Find the length of the hypotenuse of a right triangle whose sides are 10 inches and 24 inches.
    (a) 5 in.
    (b) 26 in.
    (c) 12 in.
    (d) 15 in.

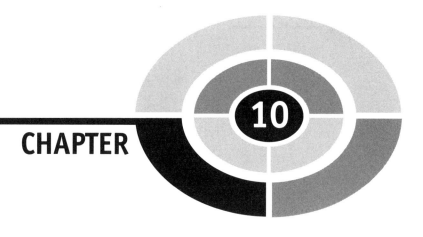

CHAPTER 10

Measurement

Basic Concepts

Many times in real-life situations, you must convert measurement units from one form to another. For example, you may need to change feet to yards, or quarts to gallons, etc. There are several mathematical methods that can be used. The easiest one to use is to multiply or divide.

To change from a larger unit to a smaller unit, **multiply** *by the conversion factor. To change from a smaller unit to a larger one,* **divide** *by the conversion unit.*

The conversion units in the tables in this chapter are arranged in descending order: i.e., the largest unit is on top, and the smallest unit is at the bottom. This chapter includes conversion units for length, weight, capacity, and time. Only the most common units of measurement are explained in this chapter. Units such as rods, nautical miles, fathoms, and barrels have been omitted.

Measures of Length

Length is measured in inches, feet, yards, and miles (Table 10.1).

> **Table 10.1   Conversion Factors for Length**
> 1 mile (mi) = 1760 yards (yd)
> 1 yard (yd) = 3 feet (ft)
> 1 foot (ft) = 12 inches (in.)

EXAMPLE
Change 5 yards to feet.

SOLUTION
Since we are going from large to small, and the conversion factor is 1 yd = 3 ft, we multiply by 3 ft:

$$5 \text{ yd} \times 3 \text{ ft} = 15 \text{ ft}$$

EXAMPLE
Change 648 inches to feet.

SOLUTION
Since we are going from small to large, and the conversion factor is 1 ft = 12 in., we divide:

$$648 \text{ in.} \div 12 \text{ in.} = 54 \text{ ft.}$$

Sometimes it is necessary to use an operation more than once to arrive at the solution.

EXAMPLE
Change 79,200 feet to miles.

SOLUTION
Since no conversion factor for changing feet to miles is given in Table 10.1, it is necessary to convert feet to yards and then convert yards to miles:

$$79,200 \text{ ft} \div 3 \text{ ft} = 26,400 \text{ yd}$$
$$26,400 \text{ yd} \div 1760 \text{ yd} = 15 \text{ miles}$$

> **Math Note:** If you know the conversion factor 1 mile = 5280 feet, it is only necessary to divide once:
> $$79,200 \text{ ft} \div 5280 \text{ ft} = 15 \text{ mi}$$

Another type of conversion has mixed units.

EXAMPLE
Change 8 feet 7 inches to inches.

SOLUTION
First change 8 feet to inches and then add 7 inches to the answer:

$$8 \text{ ft} \times 12 \text{ in.} = 96 \text{ in.}$$

$$96 \text{ in.} + 7 \text{ in.} = 103 \text{ in.}$$

EXAMPLE
Change 35 feet to yards and write the remainder in feet.

SOLUTION

$$
\begin{array}{r}
11 \\
3\overline{)35} \\
\underline{3} \\
5 \\
\underline{3} \\
2
\end{array}
$$

Hence, the answer is 11 yd 2 ft.

Math Note: The answer in the previous problem could also be written as $11\frac{2}{3}$ yards.

EXAMPLE
Change 8 yards 15 inches to yards.

SOLUTION

$$8 \text{ yd } 15 \text{ in.} = 8 \text{ yd} + 15 \text{ in.}$$

$$= 8 \text{ yd} + \frac{15}{36} \text{ in.} \qquad \text{Change inches to yards}$$

$$= 8\frac{15}{36} \text{ yd}$$

$$= 8\frac{5}{12} \text{ yd}$$

PRACTICE
Change:

1. 16 ft to inches
2. 189 ft to yards
3. 9 ft 6 in. to inches

4. 19 yd to inches
5. 24,640 yd to miles
6. 14 mi to feet
7. 31,680 ft to miles
8. 6000 yd to miles (write the remainder in yards)
9. 43 ft 9 in. to feet
10. 19 yd 8 ft 5 in. to inches

ANSWERS

1. 192 in.
2. 63 yd
3. 114 in.
4. 684 in.
5. 14 mi
6. 73,920 ft
7. 6 mi
8. 3 mi 720 yd
9. $43\frac{3}{4}$ ft
10. 785 in.

Measures of Weight

Weight is measured in ounces, pounds, and tons. Use the conversion factors in Table 10.2 and follow the rules stated in the Basic Concepts.

Table 10.2   Conversion Factors for Weight
1 ton (T) = 2000 pounds (lb)
1 pound (lb) = 16 ounces (oz)

EXAMPLE
Change 9 pounds to ounces.

SOLUTION

$$9 \text{ lb} \times 16 \text{ oz} = 144 \text{ oz}$$

EXAMPLE
Change 11,000 pounds to tons.

SOLUTION

$$11,000 \div 2000 = 5.5 \text{ T}$$

EXAMPLE
Change 2 tons to ounces.

SOLUTION

$$2 \text{ T} \times 2000 \text{ lb} = 4000 \text{ lb}$$
$$4000 \text{ lb} \times 16 \text{ oz} = 64,000 \text{ oz}$$

EXAMPLE
Change 14 pounds 3 ounces to ounces.

SOLUTION
14 lb × 16 oz = 224 oz
224 oz + 3 oz = 227 oz

EXAMPLE
Change 70 ounces to pounds and write the remainder in ounces.

SOLUTION

$$70 \text{ oz} \div 16 \text{ oz} = 16\overline{)70} \begin{array}{c} 4 \\ \end{array}$$
$$\begin{array}{r} 64 \\ \hline 6 \end{array}$$

Hence, the answer is 4 lb 6 oz.

PRACTICE
Change:

1. 13 lb to ounces
2. 26,000 lb to tons
3. 3 lb 14 oz to ounces
4. 56 oz to pounds (write the remainder in ounces)
5. 4.5 T to ounces
6. 3000 oz to pounds
7. 5 oz to pounds

8. 1 T 600 lb to pounds
9. 1,440,000 oz to tons
10. $2\frac{3}{4}$ lb to ounces

ANSWERS
1. 208 oz
2. 13 T
3. 62 oz
4. 3 lb 8 oz
5. 144,000 oz
6. 187.5 lb
7. 0.3125 lb
8. 2600 lb
9. 45 T
10. 44 oz

Measures of Capacity

Liquids are measured in ounces, pints, quarts, and gallons (Table 10.3).

Table 10.3   Conversion Factors for Capacity (Liquid)
1 gallon (gal) = 4 quarts (qt)
1 quart (qt) = 2 pints (pt)
1 pint (pt) = 16 ounces (oz)

Math Note: You should be careful to distinguish between a liquid ounce, which is a measure of capacity, and an ounce, which is a measure of weight. Both measures are called ounces, but they are different.

EXAMPLE
Change 5 gallons to quarts.

SOLUTION

$$5 \text{ gal} \times 4 \text{ qt} = 20 \text{ qt}$$

EXAMPLE
Change 7 pints to quarts.

SOLUTION

$$7 \text{ pt} \div 2 \text{ pt} = 3.5 \text{ qt}$$

EXAMPLE
Change 3 pints 10 ounces to pints.

SOLUTION

$$3 \text{ pt } 10 \text{ oz} = 3 \text{ pt } + 10 \text{ oz}$$
$$= 3 \text{ pt } + \frac{10}{16} \text{ pt}$$
$$= 3\frac{10}{16} \text{ pt}$$
$$= 3\frac{5}{8} \text{ pt}$$

EXAMPLE
Change 72 pints to gallons.

SOLUTION

$$72 \text{ pt} \div 2 \text{ pt} = 36 \text{ qt}$$
$$36 \text{ qt} \div 4 \text{ qt} = 9 \text{ gal}$$

PRACTICE
Change:

1. 5 pt to ounces
2. 240 oz to pints
3. 7 pt 9 oz to ounces
4. 15 qt to pints
5. 6 gal to ounces
6. 45 oz to pints (write the remainder in ounces)
7. 9 qt 3 pt to quarts
8. 55 pt to gallons (write the remainder in quarts and pints)
9. 3.5 gal to pt
10. 16 oz to quarts

ANSWERS

1. 80 oz
2. 15 pt
3. 121 oz
4. 30 pt
5. 768 oz
6. 2 pt 13 oz
7. 10.5 qt
8. 6 gal 3 qt 1 pt
9. 28 pt
10. 0.5 qt

Measures of Time

Time is measured in years, weeks, days, hours, minutes, and seconds (Table 10.4).

Table 10.4   Conversion Factors for Time

1 year (yr) = 12 months (mo)
             = 52 weeks (wk)
             = 365 days (da)*
1 week (wk) = 7 days
1 day (da) = 24 hours (hr)
1 hour (hr) = 60 minutes (min)
1 minute (min) = 60 seconds (sec)

*Leap years will not be used.

EXAMPLE
Change 5 years to months.

SOLUTION

$$5 \text{ yr} \times 12 \text{ mo} = 60 \text{ mo}$$

EXAMPLE
Change 3212 days to years.

SOLUTION

$$3212 \text{ da} \div 365 \text{ da} = 8.8 \text{ yr}$$

EXAMPLE

Change 54 days to weeks (write the remainder in days).

SOLUTION

$$
\begin{array}{r}
7 \\
7\overline{)54} \\
\underline{49} \\
5
\end{array}
$$

54 days is equivalent to 7 weeks and 5 days.

PRACTICE

Change:

1. 9120 min to hours
2. 45 wk to days
3. 8 hr 15 min to minutes
4. 67 mo to years (write the remainder in months)
5. 48 min to hours
6. 52 hr 34 min to seconds
7. 3.5 yr to weeks
8. 3 hr 45 min to hours
9. 10 min 45 sec to seconds
10. 3 yr to days

ANSWERS

1. 152 hr
2. 315 da
3. 495 min
4. 5 yr 7 mo
5. $\dfrac{4}{5}$ or 0.8 hr
6. 189,240 sec
7. 182 wk
8. 3.75 hr or $3\frac{3}{4}$ hr
9. 645 sec
10. 1095 da

Word Problems

The following procedure can be used to solve word problems involving measurement:

- *Step 1   Read the problem and decide what you are being asked to find.*
- *Step 2   Select the correct conversion factor or factors.*
- *Step 3   Multiply or divide.*
- *Step 4   Complete any additional steps necessary.*

EXAMPLE
Find the cost of 24 feet of rope if it costs $0.59 per yard.

SOLUTION
Change 24 ft to yards:

$$24 \text{ ft} \div 3 \text{ ft} = 8 \text{ yd}$$

Find the cost of 8 yd:

$$\$0.59 \times 8 = \$4.72$$

Hence, 24 feet of rope will cost $4.72.

EXAMPLE
If a gallon of oil costs $4.95 and a quart of oil costs $1.79, how much is saved by buying the oil in gallons instead of quarts if a person needs 3 gallons of oil?

SOLUTION
Cost of 3 gallons of oil

$$\$4.95 \times 3 = \$14.85$$

Change 3 gallons to quarts:

$$3 \text{ gal} \times 4 \text{ qt} = 12 \text{ qt}$$

Cost of 12 qt of oil

$$\$1.79 \times 12 = \$21.48$$

Subtract:

$$\$21.48 - \$14.85 = \$6.63$$

Hence, a person can save $6.63 by buying the oil in gallons instead of quarts.

PRACTICE

1. If a family uses one quart of milk per day, how many gallons are used in one year?
2. How many 6-ounce boxes of candy can be filled from a 3-pound box of candy?
3. How many cubic feet of salt water are in a container that weighs one ton? One cubic foot of water weighs 64 pounds (ignore the weight of the container).
4. If the speed of light is 186,000 miles per second, how long (in hours) does it take the light from the sun to reach the planet Pluto, if it is 2,763,000,000 miles from the sun?
5. It takes 6 inches of ribbon to make a decorative identification badge. Find the cost of making 30 badges, if ribbon costs $0.39 per yard?

ANSWERS

1. 91.25 gal
2. 8 boxes
3. 31.25 gal
4. 4.13 hr (rounded)
5. $1.95

Quiz

1. Change 9 feet to inches.
    (a) 27 in.
    (b) 90 in.
    (c) 108 in.
    (d) 180 in.

2. Change 75 feet to yards.
    (a) 5 yd
    (b) 25 yd
    (c) 150 yd
    (d) 225 yd

3. Change 33,264 feet to miles.
    (a) 6.3 mi
    (b) 18.9 mi
    (c) 12.6 mi
    (d) 9.45 mi

4. Change 5 yards 2 feet to inches.
    (a) 17 in.
    (b) 182 in.
    (c) 196 in.
    (d) 204 in.

5. Change 123.7 yards to miles.
    (a) 0.23
    (b) 0.56 mi
    (c) 0.7 mi
    (d) 0.14 mi

6. Change 16 yards, 6 feet, 10 inches to inches.
    (a) 658 in.
    (b) 130 in.
    (c) 576 in.
    (d) 340 in.

7. Change 86 pounds to ounces.
    (a) 344 oz
    (b) 688 oz
    (c) 1032 oz
    (d) 1376 oz

8. Change 10,800 pounds to tons.
    (a) 5 T
    (b) 5.4 T
    (c) 54 T
    (d) 540 T

9. Change 43 ounces to pounds and write the remainder in ounces.
    (a) 1 lb 11 oz
    (b) 3 lb 9 oz
    (c) 2 lb 9 oz
    (d) 2 lb 11 oz

10. Change 6 tons to ounces.
    (a) 12,000 oz
    (b) 192,000 oz
    (c) 96,000 oz
    (d) 36,000 oz

11. Change $6\frac{7}{8}$ pounds to ounces.
    (a) 220 oz

(b) 192 oz

(c) 96 oz

(d) 110 oz

12. Change 9000 pounds to tons.
  (a) 4 T
  (b) 9 T
  (c) 10 T
  (d) 4.5 T

13. Change 4.5 gallons to ounces.
  (a) 144 oz
  (b) 288 oz
  (c) 576 oz
  (d) 10,368 oz

14. Change 14 quarts to pints.
  (a) 7 pt
  (b) 28 pt
  (c) 84 pt
  (d) 3.5 pt

15. Change 12 pints 3 ounces to ounces.
  (a) 96 oz
  (b) 99 oz
  (c) 192 oz
  (d) 195 oz

16. Change 102 months to years.
  (a) 4.25 yr
  (b) 6 yr
  (c) 8.5 yr
  (d) 16 yr

17. Change 45 seconds to minutes.
  (a) 0.75 min
  (b) 270 min
  (c) 0.25 min
  (d) 2700 min

18. Change 50 days into weeks and write the remainder in days.
  (a) 6 wk 2 da
  (b) 5 wk 4 da
  (c) 7 wk 1 da
  (d) 9 wk 3 da

19. How high in miles is Mt. McKinley if it is 20,320 feet high?
    (a) about 5.4 mi
    (b) about 11.5 mi
    (c) about 3.8 mi
    (d) about 6.4 mi

20. Find the cost per ounce of milk if a gallon of milk sells for $2.56.
    (a) $0.06
    (b) $0.04
    (c) $0.02
    (d) $0.08

CHAPTER 11

Graphing

The Rectangular Coordinate Plane

In Chapter 9 some concepts of geometry were explained. This chapter explains the concepts of combining algebra and geometry using the **rectangular coordinate plane**.

The rectangular coordinate plane uses two perpendicular number lines called **axes**. The horizontal axis is called the **x axis**. The vertical axis is called the **y axis**. The intersection of the axes is called the **origin**. The axes divide the plane into four **quadrants**, called I, II, III, and IV (see Fig. 11-1).

Plotting Points

Each point on the plane can be located by its **coordinates**. The coordinates give the horizontal and vertical distances from the y axis and x axis, respectively. The distances are called the **x coordinate (abscissa)** and the **y coordinate (ordinate)**, and they are written as an **ordered pair** (x, y). For example, a point with coordinates (2, 3) is located two units to the right of the y axis and 3 units above the x axis (see Fig. 11-2).

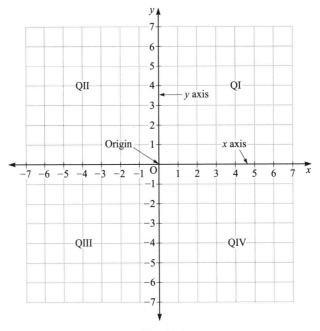

Fig. 11-1.

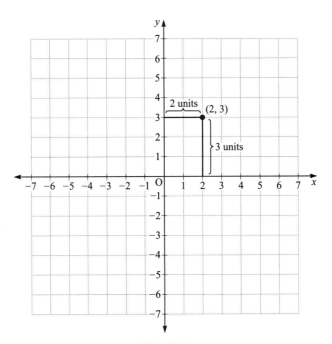

Fig. 11-2.

The point whose coordinates are (2, 3) is located in the first quadrant or QI, since both coordinates are positive. The point whose coordinates are (−4, 1) is located in the second quadrant or QII, since the x coordinate is negative and the y coordinate is positive. The point whose coordinates are (−3, −5) is in the third quadrant or QIII, since both coordinates are negative. The point (3, −1) is located in the fourth quadrant or QIV, since the x coordinate is positive and the y coordinate is negative (see Fig. 11-3).

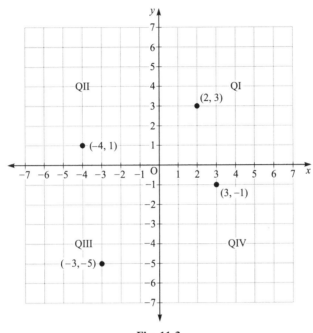

Fig. 11-3.

The coordinates of the origin are (0, 0).

Any point whose y coordinate is zero is located on the x axis. For example, point P, whose coordinates are (−3, 0), is located on the x axis 3 units to the left of the y axis. Any point whose x coordinate is zero is located on the y axis. For example, the point Q, whose coordinates are (0, 4), is located on the y axis four units above the x axis (see Fig. 11-4).

EXAMPLE
Give the coordinates of each point shown in Fig. 11-5.

SOLUTION
A (−2, 4); B (3, 1); C (−1, −5); D (6, −2); E (2, 0); F (0, −4).

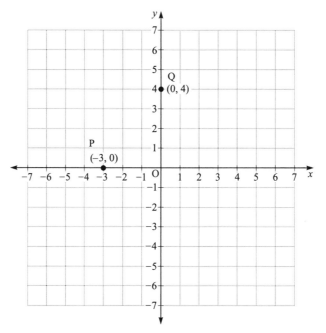

Fig. 11-4.

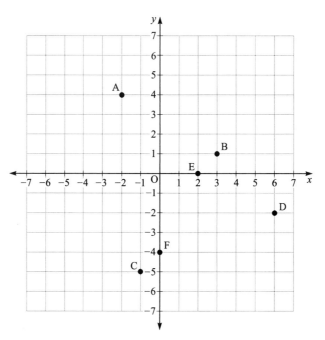

Fig. 11-5.

PRACTICE

Give the coordinates of each point shown in Fig. 11-6.

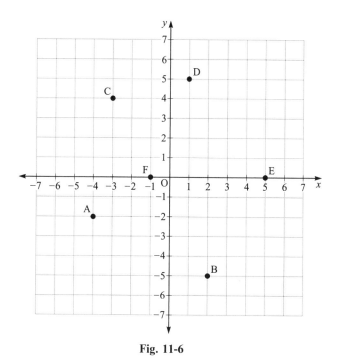

Fig. 11-6

ANSWERS

1. A $(-4, -2)$
2. B $(2, -5)$
3. C $(-3, 4)$
4. D $(1, 5)$
5. E $(5, 0)$
6. F $(0, -1)$

Linear Equations

An equation such as $x + 2y = 6$ is called a **linear equation** in two variables. A solution to the equation is a set of ordered pairs (x, y) such that when the values are substituted for the variables in the equation, a closed true equation results. For example (2, 2) is a solution, since

$$x + 2y = 6$$
$$2 + 2(2) = 6$$
$$2 + 4 = 6$$
$$6 = 6$$

Another solution to the equation $x + 2y = 6$ is $(4, 1)$, since

$$x + 2y = 6$$
$$4 + 2(1) = 6$$
$$4 + 2 = 6$$
$$6 = 6$$

The linear equation $x + 2y = 6$ actually has an *infinite* number of solutions. For any value of x, a corresponding value of y can be found by substituting the value for x into the equation and solving it for y. For example, if $x = 8$, then

$$x + 2y = 6$$
$$8 + 2y = 6$$
$$8 - 8 + 2y = 6 - 8$$
$$2y = -2$$
$$\frac{2y}{2} = \frac{-2}{2}$$
$$y = -1$$

When $x = 8$, $y = -1$ and the ordered pair $(8, -1)$ is a solution, too.

EXAMPLE
Find a solution to the equation $3x - y = 10$.

SOLUTION
Select any value for x, say $x = 4$, substitute in the equation, and then solve for y:

$$3x - y = 10$$
$$3(4) - y = 10$$
$$12 - y = 10$$
$$12 - 12 - y = 10 - 12$$
$$-y = -2$$
$$\frac{-1y}{-1} = \frac{-2}{-1}$$
$$y = 2$$

Hence, (4, 2) is a solution to $3x - y = 10$.

Math Note: The solution can be checked by substituting both values in the equation:

$$3x - y = 10$$
$$3(4) - 2 = 10$$
$$12 - 2 = 10$$
$$10 = 10$$

EXAMPLE

Given the equation $5x - 2y = 16$, find y when $x = 4$.

SOLUTION

Substitute 4 in the equation and solve for y:

$$5x - 2y = 16$$
$$5(4) - 2y = 16$$
$$20 - 2y = 16$$
$$20 - 20 - 2y = 16 - 20$$
$$-2y = -4$$
$$\frac{-2y}{-2} = \frac{-4}{-2}$$
$$y = 2$$

Hence, when $x = 4$, $y = 2$. The ordered pair (4, 2) is a solution for $5x - 2y = 16$.

PRACTICE

1. Find y when $x = 5$ for $3x + 4y = 27$
2. Find y when $x = -1$ for $x + 3y = 14$
3. Find y when $x = 0$ for $5x + 2y = 20$
4. Find y when $x = 3$ for $-2x + y = 10$
5. Find y when $x = -4$ for $3x - 2y = 10$

ANSWERS

1. $y = 3$
2. $y = 5$
3. $y = 10$
4. $y = 16$
5. $y = -11$

Graphing Lines

Equations of the form ax + by = c where a, b, and c are real numbers are called **linear equations in two variables** and their graphs are straight lines.

In order to graph a linear equation, find the coordinates of two points on the line (i.e., solutions), and then plot the points and draw the line through the two points.

EXAMPLE

Draw the graph of x + y = 6.

SOLUTION

Select two values for x and find the corresponding y values:

Let x = 3   Let x = 5
x + y = 6   x + y = 6
3 + y = 6   5 + y = 6
3 − 3 + y = 6 − 3   5 − 5 + y = 6 − 5
y = 3   y = 1
(3, 3)   (5, 1)

Plot the points and draw a line through the two points (see Fig. 11-7).

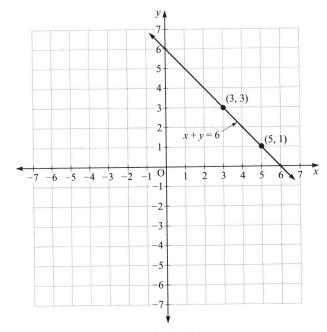

Fig. 11-7.

EXAMPLE
Draw the graph of $2x - y = 8$.

SOLUTION
Find the coordinates of two points on the line:

Let $x = 1$ Let $x = 5$
$2x - y = 8$ $2x - y = 8$
$2(1) - y = 8$ $2(5) - y = 8$
$2 - y = 8$ $10 - y = 8$
$2 - 2 - y = 8 - 2$ $10 - 10 - y = 8 - 10$
$-y = 6$ $-y = -2$
$\dfrac{-y}{-1} = \dfrac{6}{-1}$ $\dfrac{-y}{-1} = \dfrac{-2}{-1}$
$y = -6$ $y = 2$
$(1, -6)$ $(5, 2)$

Plot the two points $(1, -6)$ and $(5, 2)$ and draw the line through these points (see Fig. 11-8).

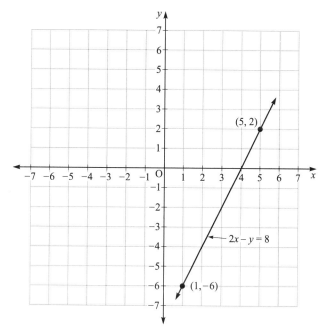

Fig. 11-8.

Math Note: When graphing lines, it is best to select three points rather than two points in case an error has been made. If an error has been made, the three points will not line up.

PRACTICE
Draw the graph of each of the following:

1. $x - y = 3$
2. $-x + 4y = 9$
3. $2x + 5y = 7$
4. $3x - y = 6$
5. $-5x + y = -11$

ANSWERS
1. See Fig. 11-9
2. See Fig. 11-10
3. See Fig. 11-11
4. See Fig. 11-12
5. See Fig. 11-13

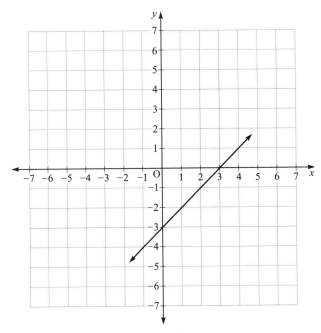

Fig. 11-9.

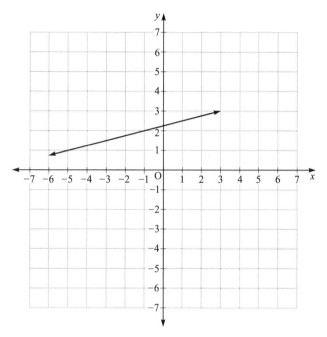

Fig. 11-10.

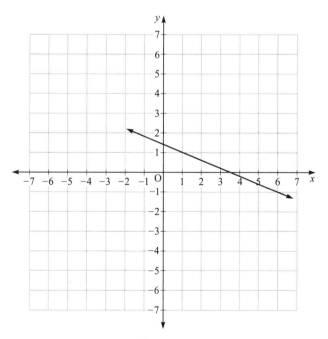

Fig. 11-11.

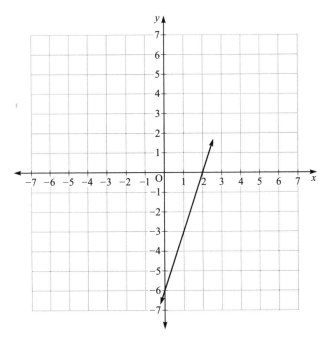

Fig. 11-12.

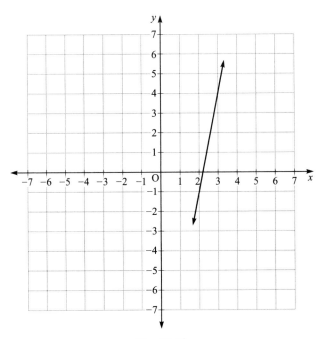

Fig. 11-13.

Horizontal and Vertical Lines

Any equation of the form x = a, where a is a real number, is the graph of a vertical line passing through the point (a, 0) on the x axis.

EXAMPLE
Graph the line x = −5.

SOLUTION
Draw a vertical line passing through the point (−5, 0) on the x axis (see Fig. 11-14).

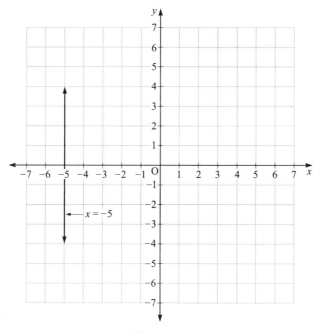

Fig. 11-14.

Any equation of the form y = b, where b is any real number, is the graph of a horizontal line passing through the point (0, b) on the y axis.

EXAMPLE
Graph the line y = 4.

SOLUTION
Draw a horizontal line through the point (0, 4) on the y axis (see Fig. 11-15).

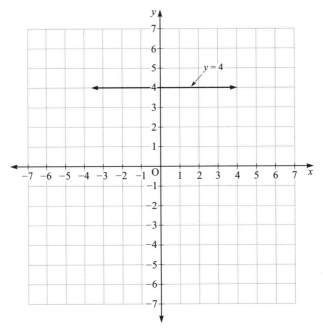

Fig. 11-15.

PRACTICE

Draw the graph of each of the following:

1. $y = 6$
2. $x = -4$
3. $y = -2$
4. $x = 3$
5. $y = 2$

ANSWERS

1. See Fig. 11-16
2. See Fig. 11-17
3. See Fig. 11-18
4. See Fig. 11-19
5. See Fig. 11-20

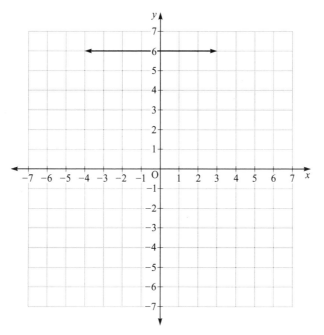

Fig. 11-16.

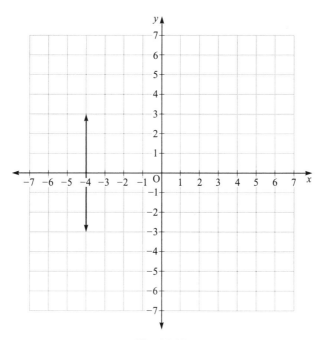

Fig. 11-17.

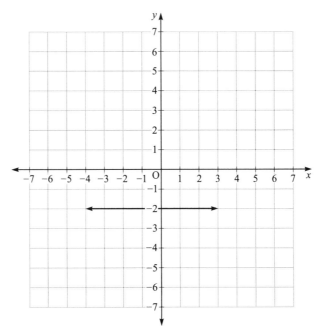

Fig. 11-18.

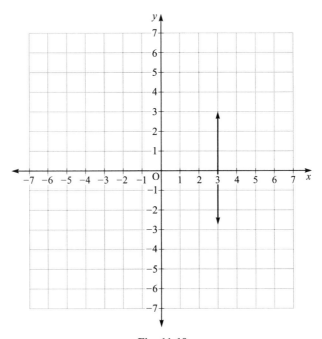

Fig. 11-19.

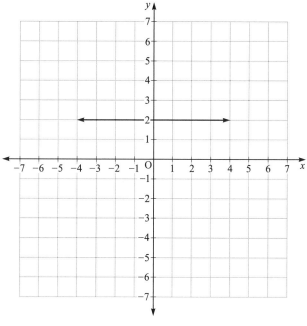

Fig. 11-20.

Intercepts

The point on the graph where a line crosses the y axis is called the **y intercept** (see Fig. 11-21).

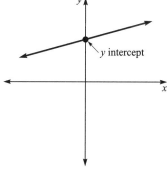

Fig. 11-21.

To find the y intercept, let x = 0 and substitute in the equation, and then solve for y.

EXAMPLE
Find the y intercept of $2x + 3y = 9$.

SOLUTION
Substitute 0 for x and solve for y:

$$2x + 3y = 9$$
$$2(0) + 3y = 9$$
$$0 + 3y = 9$$
$$3y = 9$$
$$\frac{3y}{3} = \frac{9}{3}$$
$$y = 3$$

Hence, the y intercept is (0, 3).

The point on the graph where a line crosses the x axis is called the **x intercept** (see Fig. 11-22).

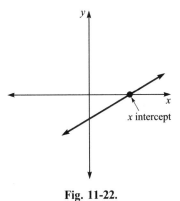

Fig. 11-22.

To find the x intercept, let $y = 0$ and substitute in the equation, and then solve for x.

EXAMPLE
Find the x intercept for $5x + 2y = 30$.

SOLUTION
Substitute 0 for y and solve for x:

$$5x + 2y = 30$$
$$5x + 2(0) = 30$$
$$5x + 0 = 30$$
$$5x = 30$$
$$\frac{5x}{5} = \frac{30}{5}$$
$$x = 6$$

Hence, the x intercept is (6, 0).

Math Note: It is easy to draw lines on a graph using the intercepts for the two points.

PRACTICE
1. Find the x intercept of $5x + 4y = 30$
2. Find the y intercept of $x - 7y = 14$
3. Find the y intercept of $2x + 8y = 24$
4. Find the x intercept of $3x - 17y = 21$
5. Find the y intercept of $x + 5y = -20$

ANSWERS
1. (6, 0)
2. (0, −2)
3. (0, 3)
4. (7, 0)
5. (0, −4)

Slope

An important concept associated with lines is called the *slope* of a line. The slope of a line is associated with the "steepness" of a line. The **slope** of a line is the ratio of the vertical change to the horizontal change of a line when going from left to right. The slope is loosely defined as the rise divided by the run (see Fig. 11-23).

The slope of a line can be found in several ways. A line going uphill from left to right has a slope that is positive. A line going downhill from left to right has a slope that is negative. The slope of a horizontal line is zero and the slope of a vertical line is *undefined* (see Fig. 11-24).

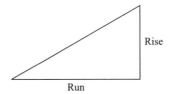

Fig. 11-23.

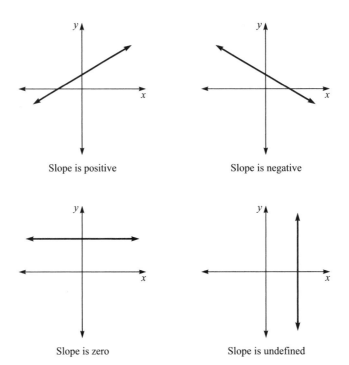

Fig. 11-24.

On the graph of a line the slope can be found by selecting two points, forming a right triangle, and then counting the number of units in the rise and run and dividing those two values (see Fig. 11-25).

A better method is to find the coordinates of two points on the line, say (x_1, y_1) *and* (x_2, y_2) *and then use the formula:*

$$slope\ (m) = \frac{y_2 - y_1}{x_2 - x_1}$$

EXAMPLE

Find the slope of a line passing through the points whose coordinates are (5, 3) and (8, 1).

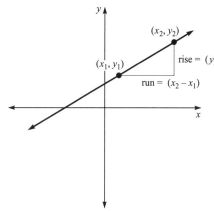

Fig. 11-25.

SOLUTION
Let $x_1 = 5$, $y_1 = 3$, and $x_2 = 8$, $y_2 = 1$, and then substitute in the slope formula:

$$m = \frac{y_2 - y_1}{x_2 - x_1}$$

$$= \frac{1 - 3}{8 - 5}$$

$$= \frac{-2}{3} = -\frac{2}{3}$$

Hence, the slope of the line is $-\frac{2}{3}$.

The slope whose equation is known can be found by selecting two points on the line then using the slope formula.

EXAMPLE
Find the slope of a line whose equation is $5x + 2y = 10$.

SOLUTION
Select two points on the line:

Let $x = 4$, then	Let $x = -2$, then
$5x + 2y = 10$	$5x + 2y = 10$
$5(4) + 2y = 10$	$5(-2) + 2y = 10$
$20 + 2y = 10$	$-10 + 2y = 10$
$20 - 20 + 2y = 10 - 20$	$-10 + 10 + 2y = 10 + 10$
$2y = -10$	$2y = 20$
$\dfrac{2y}{2} = \dfrac{-10}{2}$	$\dfrac{2y}{2} = \dfrac{20}{2}$
$y = -5$	$y = 10$
$(4, -5)$	$(-2, 10)$

Let $x_1 = 4, y_1 = -5$ and $x_2 = -2, y_2 = 10$. Now substitute in the slope formula:

$$m = \frac{y_2 - y_1}{x_2 - x_1}$$

$$m = \frac{10 - (-5)}{-2 - 4}$$

$$= \frac{15}{-6}$$

$$= -\frac{5}{2}$$

The slope of the line $5x + 2y = 10$ is $-\frac{5}{2}$.

Another way to find the slope of a line is to solve the equation for y in terms of x. The coefficient of x will be the slope.

EXAMPLE

Find the slope of a line whose equation is $5x + 2y = 10$.

SOLUTION

Solve the equation for y, as shown:

$$5x + 2y = 10$$

$$5x - 5x + 2y = -5x + 10$$

$$2y = -5x + 10$$

$$\frac{2y}{2} = \frac{-5x}{2} + \frac{10}{2}$$

$$y = -\frac{5}{2}x + 5$$

Hence, the slope is $-\frac{5}{2}$, which is the result found in the previous example.

Math Note: An equation in the form of y in terms of x is called the slope–intercept form and in general is written as $y = mx + b$, where m is the slope and b is the y intercept.

PRACTICE
1. Find the slope of the line containing the two points $(-3, 4)$ and $(6, 8)$.
2. Find the slope of the line containing the two points $(9, 3)$ and $(-4, -2)$.
3. Find the slope of the line whose equation is $5x - 3y = 8$.
4. Find the slope of the line whose equation is $-2x + 4y = 9$.
5. Find the slope of the line whose equation is $x + 7y = 10$.

ANSWERS
1. $\dfrac{4}{9}$

2. $\dfrac{5}{13}$

3. $\dfrac{5}{3}$

4. $\dfrac{1}{2}$

5. $-\dfrac{1}{7}$

Solving a System of Linear Equations

Two linear equations of the form $ax + by = c$, where a, b, and c are real numbers, is called a **system** of linear equations. When the two lines intersect, the coordinates of the point of intersection are called the **solution** of the system. The system is then said to be **independent** and **consistent**. When the lines are **parallel**, there is no point of intersection; hence, there is no solution for the system. In this case the system is said to be **inconsistent**. When the two lines coincide, every point on the line is a solution. The system is said to be **dependent** (see Fig. 11-26).

To find a solution to a system of linear equations, plot the graphs for the lines and find the point of intersection.

EXAMPLE
Find the solution of:

$$x + y = 8$$
$$x - y = 2$$

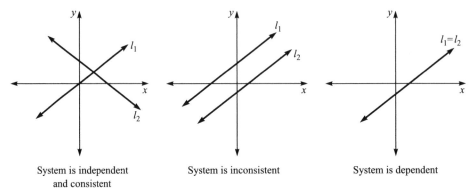

System is independent
and consistent

System is inconsistent

System is dependent

Fig. 11-26.

SOLUTION

Find two points on each line. For $x + y = 8$, use (7, 1) and (3, 5). For $x - y = 2$, use (3, 1) and (6, 4). Graph both lines and then find the point of intersection (see Fig. 11-27).

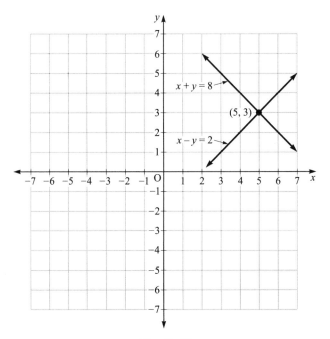

Fig. 11-27.

The point of intersection is (5, 3).

Math Note: To check, substitute the values of x and y for the solution in both equations and see if they are closed true equations.

EXAMPLE

Find the solution of:

$$3x - y = 8$$
$$x + 2y = 5$$

SOLUTION

Find two points on each line. For $3x - y = 8$, use $(2, -2)$ and $(4, 4)$. For $x + 2y = 5$, use $(7, -1)$ and $(1, 2)$. Graph both lines and then find the point of intersection (see Fig. 11-28).

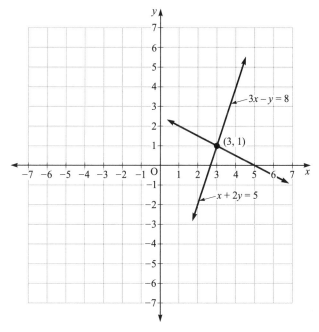

Fig. 11-28.

The point of intersection is $(3, 1)$.

PRACTICE

Find the solution for each of the following system of equations:

1. $x - y = 4$
   $x + y = 10$

2. $2x + 3y = 12$
    $x + 4y = 11$
3. $4x + y = 10$
    $2x + y = 6$
4. $x - 5y = 6$
    $5x - y = 6$
5. $x + 3y = -7$
    $6x - y = 15$

ANSWERS
1. $(7, 3)$
2. $(3, 2)$
3. $(2, 2)$
4. $(1, -1)$
5. $(2, -3)$

Quiz

Use Fig. 11-29 to answer questions 1–5.

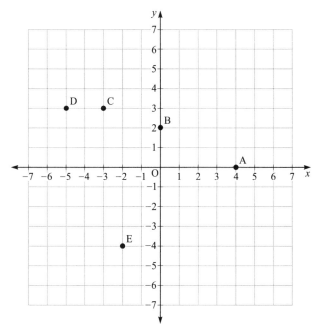

Fig. 11-29.

1. The point whose coordinates are $(-5, 3)$ is:
   (a) Point A
   (b) Point B
   (c) Point D
   (d) Point E

2. The point whose coordinates are $(4, 0)$ is:
   (a) Point A
   (b) Point B
   (c) Point C
   (d) Point E

3. The point whose coordinates are $(-3, 3)$ is:
   (a) Point B
   (b) Point C
   (c) Point D
   (d) Point E

4. The point whose coordinates are $(0, 2)$ is:
   (a) Point A
   (b) Point B
   (c) Point C
   (d) Point D

5. The point whose coordinates are $(-2, -4)$ is:
   (a) Point A
   (b) Point B
   (c) Point C
   (d) Point E

Use Figure 11-30 to answer questions 6–9.

6. The line whose equation is $2x + 3y = 6$ is:
   (a) line l
   (b) line m
   (c) line n
   (d) line o

7. The line whose equation is $x = 3$ is:
   (a) line l
   (b) line m
   (c) line n
   (d) line o

8. The line whose equation is $x - 2y = 4$ is:

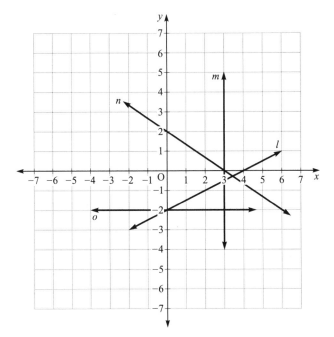

Fig. 11-30.

    (a) line l
    (b) line m
    (c) line n
    (d) line o

9. The line whose equation is $y = -2$ is:
    (a) line l
    (b) line m
    (c) line n
    (d) line o

10. Which is a solution to $5x - y = 12$?
    (a) $(2, 2)$
    (b) $(2, -12)$
    (c) $(3, 3)$
    (d) $(0, 12)$

11. Which is a solution to $-2x + y = 5$?
    (a) $(1, 3)$
    (b) $(3, -1)$
    (c) $(0, 5)$
    (d) $(2, 1)$

12. Find y when x = −3 for 2x + y = −8.
    (a) 2
    (b) 1
    (c) −2
    (d) −4

13. Find x when y = −3 for 5x + 3y = 16.
    (a) x = −5
    (b) x = 2
    (c) x = −2
    (d) x = 5

14. Find the x intercept for −7x + 2y = 14.
    (a) (−2, 0)
    (b) (0, 7)
    (c) (2, 7)
    (d) (−2, 7)

15. Find the y intercept for 3x − 8y = 24.
    (a) (8, 0)
    (b) (0, −3)
    (c) (8, 3)
    (d) (0, 3)

16. Find the slope of a line containing the two points whose coordinates are (−1, 6) and (3, 8).

    (a) $-\dfrac{2}{7}$

    (b) $\dfrac{1}{2}$

    (c) $\dfrac{2}{3}$

    (d) $\dfrac{7}{2}$

17. Find the slope of the line −6x + 3y = 10.

    (a) $-\dfrac{1}{2}$

    (b) −2

    (c) 2

(d) $\dfrac{10}{8}$

18. The slope of a vertical line is:
    (a) 0
    (b) 1
    (c) undefined
    (d) −1

19. When two lines are parallel, the system is said to be:
    (a) consistent
    (b) undefined
    (c) dependent
    (d) inconsistent

20. Which point is a solution for:

$$5x - y = 11$$
$$2x + y = 10$$

(a) (5, 0)
(b) (−2, 2)
(c) (3, 4)
(d) (10, −1)

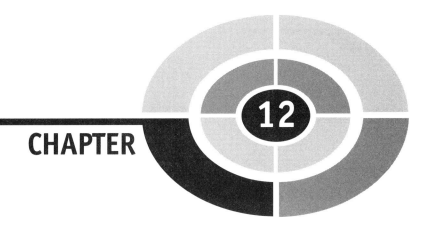

CHAPTER 12

Operations with Monomials and Polynomials

Monomials and Polynomials

Recall from Chapter 7 that an **algebraic expression** consists of variables (letters), constants (numbers), operation signs, and grouping symbols. Also recall that a **term** of an algebraic expression consists of a number, variable, or a product or quotient of numbers and variables. The terms of an algebraic expression are connected by + or − signs. An algebraic expression consisting only of the four operations (addition, subtraction, multiplication, and division), and which has no variable in the denominator of a term, is called a **polynomial**. If the expression has one term, it is called a **monomial**. If the expression has two terms, it is called a **binomial**. If the expression has three terms, it is called a **trinomial**.

Monomials:	$3x^2$	$-5y$	$2x^3y^2$
Binomials:	$x + 2y$	$3x^2 + x$	$-2y + z$
Trinomials	$3x^2 + 4x - 1$		$2x - 3y + 5$

Addition of Polynomials

Recall in Chapter 7 that only like terms can be added or subtracted. For example, $5x + 6x = 11x$. Unlike terms cannot be added or subtracted.

To add two or more polynomials, add like terms.

EXAMPLE
Add $(3x^2 + 2x - 5) + (5x^2 - 7x + 3)$.

SOLUTION

$$
\begin{aligned}
(3x^2 + 2x - 5) + (5x^2 - 7x + 3) &= 3x^2 + 2x - 5 + 5x^2 - 7x + 3 \\
&= (3x^2 + 5x^2) + (2x - 7x) + (-5 + 3) \\
&= 8x^2 - 5x - 2
\end{aligned}
$$

> **Math Note:** Recall that when the numerical coefficient is one, it is not written: i.e., $xy = 1xy$. Likewise $-xy = -1xy$.

EXAMPLE
Add $(2x + y) + (2x - 5) + (3y - 7)$.

SOLUTION

$$
\begin{aligned}
(2x + y) + (2x - 5) + (3y - 7) &= 2x + y + 2x - 5 + 3y - 7 \\
&= (2x + 2x) + (y + 3y) + (-5 - 7) \\
&= 4x + 4y - 12
\end{aligned}
$$

> **Math Note:** If the sum of two like terms is 0, do not write the term in the sum.

PRACTICE
Add:

1. $(6x^2 + 3x - 8) + (2x^2 - 5x - 9)$
2. $(3x + 2y - 2) + (4x - 2y + 8)$
3. $(2x + 5) + (7x - 10) + (3x - 4) + (4x - 6)$
4. $(3a^2b^2 - 2ab + 5) + (a^2b^2 + 2ab + 4)$
5. $(9m + 5n - 2) + (4n - 8) + (6m + 7)$

ANSWERS

1. $8x^2 - 2x - 17$
2. $7x + 6$
3. $16x - 15$
4. $4a^2b^2 + 9$
5. $15m + 9n - 3$

Subtraction of Polynomials

In Chapter 2 you learned that whenever you subtract integers in algebra, you add the opposite. For example, $-10 - (-6) = -10 + 6 = -4$. To find the opposite of an integer (except 0), we change its sign. To find the opposite of a monomial, change the sign of the numerical coefficient: e.g., the opposite of $-7xy$ is $7xy$ and the opposite of $8x^2$ is $-8x^2$.

To find the opposite of a polynomial, change the signs of every term of the polynomial: e.g., the opposite of $6x^2 - 3x + 2$ is $-(6x^2 - 3x + 2)$ or $-6x^2 + 3x - 2$.

To subtract two polynomials, add the opposite of the polynomial being subtracted.

EXAMPLE
Subtract $(9x^2 + 3x - 2) - (6x^2 + 5x - 8)$.

SOLUTION

$$(9x^2 + 3x - 2) - (6x^2 + 5x - 8) = 9x^2 + 3x - 2 - 6x^2 - 5x + 8$$
$$= (9x^2 - 6x^2) + (3x - 5x) + (-2 + 8)$$
$$= 3x^2 + (-2x) + 6$$
$$= 3x^2 - 2x + 6$$

EXAMPLE
Subtract $(2a - 3b - c) - (5a - 6b - c)$.

SOLUTION

$$(2a - 3b - c) - (5a - 6b - c) = 2a - 3b - c - 5a + 6b + c$$
$$= (2a - 5a) + (-3b + 6b) + (-c + c)$$
$$= -3a + 3b$$

PRACTICE
Subtract:

1. $(6x - 3) - (x + 8)$
2. $(2x^2 + 5x - 3) - (4x^2 + 5x + 5)$
3. $(3a - 2b + 6) - (-5a - 3b + 4)$
4. $(13r - 2s) - (15s + 3r)$
5. $(y^2 + 6y - 10) - (5 + 8y - y^2)$

ANSWERS

1. $5x - 11$
2. $-2x^2 - 8$
3. $8a + b + 2$
4. $10r - 17s$
5. $2y^2 - 2y - 15$

Multiplication of Monomials

When two variables with the same base are multiplied, the exponents of the variables are added to get the product. For example,

$$x^3 \cdot x^4 = x \cdot x \cdot x \cdot x \cdot x \cdot x \cdot x = x^{3+4} = x^7$$

In general,

$$x^m \cdot x^n = x^{m+n}$$

EXAMPLE
Multiply $y^5 \cdot y^6$.

SOLUTION

$$y^5 \cdot y^6 = y^{5+6} = y^{11}$$

To multiply two monomials, multiply the numerical coefficients (numbers) and add the exponents of the same variables (letters).

EXAMPLE
Multiply $5x^3y^2 \cdot 3x^2y^4$.

SOLUTION

$$5x^3y^2 \cdot 3x^2y^4 = 5 \cdot 3 \cdot x^3 \cdot x^2 \cdot y^2 \cdot y^4$$
$$= 15x^5 \cdot y^6$$

EXAMPLE
Multiply $-3m^5n^4 \cdot 6m^2 \cdot n^6$.

SOLUTION

$$-3m^5n^4 \cdot 6m^2 \cdot n^6 = -3 \cdot 6 \cdot m^5 \cdot m^2 \cdot n^4 \cdot n^6$$
$$= -18m^7n^{10}$$

Math Note: Recall that $x = x^1$.

EXAMPLE
Multiply $-xy \cdot 2x^2y^3$.

SOLUTION

$$-xy \cdot 2x^2y^3 = -1xy \cdot 2x^2y^3$$
$$= -1 \cdot 2 \cdot x \cdot x^2 \cdot y \cdot y^3$$
$$= -2x^3y^4$$

PRACTICE
Multiply:

1. $2xy \cdot 5xy$
2. $-8a^2 \cdot 2a^3b^2$
3. $4x^3y \cdot 3x^2y^2 \cdot 5xy$

4. $-9ab^4 \cdot 6ab^2 \cdot 3ab$
5. $7xy \cdot 3x^2z \cdot (-2yz^2)$

ANSWERS

1. $10x^2y^2$
2. $-16a^5b^2$
3. $60x^6y^4$
4. $-162a^3b^7$
5. $-42x^3y^2z^3$

Raising a Monomial to a Power

When a variable is raised to a power, the exponent of the variable is multiplied by the power. For example,

$$(x^2)^3 = x^2 \cdot x^2 \cdot x^2 = x^{2+2+2} = x^6$$

or

$$(x^2)^3 = x^{2 \cdot 3} = x^6$$

In general $(x^m)^n = x^{m \cdot n}$. To raise a monomial to a power, raise the numerical coefficient to the power and multiply the exponents of the variables by the power.

EXAMPLE
Find $(5x^3y^4)^2$.

SOLUTION

$$(5x^3y^4)^2 = 5^2 \cdot x^{3 \cdot 2} y^{4 \cdot 2}$$
$$= 25x^6y^8$$

EXAMPLE
Find $(-6a^4b^5)^3$.

SOLUTION

$$(-6a^4b^5)^3 = (-6)^3 a^{4 \cdot 3} b^{5 \cdot 3}$$
$$= -216a^{12}b^{15}$$

EXAMPLE
Find $(-xy^3)^4$.

SOLUTION

$$(-xy^3)^4 = (-1xy^3)^4$$
$$= (-1)^4 x^{1 \cdot 4} y^{3 \cdot 4}$$
$$= 1x^4 y^{12}$$
$$= x^4 y^{12}$$

PRACTICE
Find each of the following:

1. $(x^4)^5$
2. $(3a)^2$
3. $(6m^2n^4)^3$
4. $(-5a^3b^2)^2$
5. $(7x^5y^2)^3$

ANSWERS

1. x^{20}
2. $9a^2$
3. $216\ m^6n^{12}$
4. $25a^6b^4$
5. $343x^{15}y^6$

Multiplication of a Polynomial by a Monomial

When a polynomial is multiplied by a monomial, the distributive property is used. Recall from Chapter 7 that the distributive property states that $a(b + c) = a \cdot b + a \cdot c$.

To multiply a polynomial by a monomial, multiply each term in the polynomial by the monomial.

EXAMPLE
Multiply x(2x + 3y).

SOLUTION

$$x(2x + 3y) = x \cdot 2x + x \cdot 3y$$
$$= 2x^2 + 3xy$$

EXAMPLE
Multiply $5x^2y(3x + 2y - 6)$

SOLUTION

$$5x^2y(3x + 2y - 6) = 5x^2y \cdot 3x + 5x^2y \cdot 2y - 5x^2y \cdot 6$$
$$= 15x^3y + 10x^2y^2 - 30x^2y$$

PRACTICE
Multiply:

1. $8(2x + 2y - 6)$
2. $4x(5x^2 - 6x + 12)$
3. $-3xy^2(2xy + 6x^2 - 10)$
4. $-c^2d^3(3c^2 + 4cd - 5d^2)$
5. $6ab(3a - 2b + 4c^2)$

ANSWERS

1. $16x + 16y - 48$
2. $20x^3 - 24x^2 + 48x$
3. $-6x^2y^3 - 18x^3y^2 + 30xy^2$
4. $-3c^4d^3 - 4c^3d^4 + 5c^2d^5$
5. $18a^2b - 12ab^2 + 24abc^2$

Multiplication of Two Binomials

In algebra, there are some special products that you will need to know. The first one is to be able to find the product of two binomials. In order to do this, you can use the distributive law twice.

EXAMPLE
Multiply (x + 4) (x + 2).

SOLUTION

Distribute (x + 4) over (x + 2) as shown:

$$(x + 4)(x + 2) = (x + 4)x + (x + 4)2.$$

Next distribute each point as shown:

$$= x \cdot x + 4 \cdot x + x \cdot 2 + 4 \cdot 2$$
$$= x^2 + 4x + 2x + 8$$

Then combine like terms:

$$= x^2 + 6x + 8$$

Hence, (x + 4)(x + 2) = x^2 + 6x + 8

EXAMPLE

Multiply (3x + 1)(x − 2).

SOLUTION

$$(3x + 1)(x - 2) = (3x + 1)x + (3x + 1)(-2)$$
$$= 3x \cdot x + 1 \cdot x + 3x(-2) + 1(-2)$$
$$= 3x^2 + x - 6x - 2$$
$$= 3x^2 - 5x - 2$$

A shortcut method for multiplying two binomials is called the FOIL method. FOIL stands for:

$$F = \text{First}$$
$$O = \text{Outer}$$
$$I = \text{Inner}$$
$$L = \text{Last}$$

EXAMPLE

Multiply (x − 7)(x + 2) using the FOIL method.

SOLUTION

First	$x \cdot x = x^2$
Outer	$x \cdot 2 = 2x$
Inner	$-7 \cdot x = -7x$
Last	$-7 \cdot 2 = -14$

Combine the like terms:

$$2x - 7x = -5x$$

Hence, $(x - 7)(x + 2) = x^2 - 5x - 14$.

PRACTICE
Multiply:

1. $(x + 6)(x + 7)$
2. $(x - 4)(x - 6)$
3. $(x + 9)(x - 11)$
4. $(x - 3)(x + 3)$
5. $(x + 4)(x + 4)$

ANSWERS

1. $x^2 + 13x + 42$
2. $x^2 - 10x + 24$
3. $x^2 - 2x - 99$
4. $x^2 - 9$
5. $x^2 + 8x + 16$

Squaring a Binomial

Another special product results from squaring a binomial. This can be done by using the two methods shown previously; however, a short-cut rule can be used. It is:

$$(a + b)^2 = a^2 + 2ab + b^2$$
$$(a - b)^2 = a^2 - 2ab + b^2$$

In words, whenever you square a binomial, square the first term and then multiply the product of the first and second terms by 2 and square the last term.

EXAMPLE
Square $(x + 8)$.

SOLUTION

$$(x + 8)^2 = x^2 + 2 \cdot x \cdot 8 + 8^2$$
$$= x^2 + 16x + 64$$

EXAMPLE

Square $(2x - 3)$.

SOLUTION

$$(2x - 3)^2 = (2x)^2 - 2 \cdot 2x \cdot 3 + (-3)^2$$
$$= 4x^2 - 12x + 9$$

EXAMPLE

Square $(3x + 1)$.

SOLUTION

$$(3x + 1)^2 = (3x)^2 + 2 \cdot 3x \cdot 1 + 1^2$$
$$= 9x^2 + 6x + 1$$

PRACTICE

1. $(x + 1)^2$
2. $(x - 6)^2$
3. $(2x + 5)^2$
4. $(4x + 7)^2$
5. $(8x - 3)^2$

ANSWERS

1. $x^2 + 2x + 1$
2. $x^2 - 12x + 36$
3. $4x^2 + 20x + 25$
4. $16x^2 + 56x + 49$
5. $64x^2 - 48x + 9$

Multiplication of Two Polynomials

Two polynomials can be multiplied by using the distributive property as many times as needed. It is the same as shown for two binomials.

EXAMPLE

Multiply $(x + 2)(x^2 + 5x - 6)$.

SOLUTION

$$(x + 2)(x^2 + 5x - 6) = (x + 2) \cdot x^2 + (x + 2) \cdot 5x + (x + 2)(-6)$$
$$= x \cdot x^2 + 2 \cdot x^2 + x \cdot 5x + 2 \cdot 5x + x(-6) + 2(-6)$$
$$= x^3 + 2x^2 + 5x^2 + 10x - 6x - 12$$
$$= x^3 + 7x^2 + 4x - 12$$

Multiplication of polynomials can be performed vertically. The process is similar to multiplying whole numbers.

EXAMPLE
Multiply $(x + 2)(x^2 + 5x - 6)$ vertically.

SOLUTION

$$
\begin{array}{r}
x^2 + 5x - 6 \\
x + 2 \\
\hline
x^3 + 5x^2 - 6x \\
2x^2 + 10x - 12 \\
\hline
x^3 + 7x^2 + 4x - 12
\end{array}
$$

Multiply the top by x

Multiply the top by 2

Add like terms

EXAMPLE
Multiply $(x^2 + 3x - 8)(3x - 5)$ vertically.

SOLUTION

$$
\begin{array}{r}
x^2 + 3x - 8 \\
3x - 5 \\
\hline
3x^3 + 9x^2 - 24x \\
-5x^2 - 15x + 40 \\
\hline
3x^3 + 4x^2 - 39x + 40
\end{array}
$$

Math Note: Be sure to place like terms under each other when multiplying vertically.

PRACTICE
Multiply:

1. $(x^2 + 3x + 2)(x + 1)$
2. $(3x^2 + 2x + 5)(x - 6)$
3. $(2x^2 + 4x - 1)(2x - 3)$
4. $(x^3 - 2x^2 + 2x + 1)(x - 5)$
5. $(x^2 + 2x + 7)(x^2 - x - 6)$

ANSWERS
1. $x^3 + 4x^2 + 5x + 2$
2. $3x^3 - 16x^2 - 7x - 30$
3. $4x^3 + 2x^2 - 14x + 3$
4. $x^4 - 7x^3 + 12x^2 - 9x - 5$
5. $x^4 + x^3 - x^2 - 19x - 42$

Division of Monomials

When one variable is divided by another variable with the same base, the exponent of the variable in the denominator is subtracted from the exponent of the variable in the numerator. For example,

$$\frac{x^6}{x^2} = \frac{\not{x} \cdot \not{x} \cdot x \cdot x \cdot x \cdot x}{\not{x} \cdot \not{x}} = x^{6-2} = x^4$$

In general,

$$\frac{x^m}{x^n} = x^{m-n}$$

EXAMPLE
Divide x^5 by x^3.

SOLUTION

$$\frac{x^5}{x^3} = x^{5-3} = x^2$$

To divide a monomial by a monomial, divide the numerical coefficients and then subtract the exponents of the same variables.

EXAMPLE
Divide $15x^5$ by $-5x^3$.

SOLUTION

$$\frac{15x^5}{-5x^3} = -3x^{5-3} = -3x^2$$

EXAMPLE
Divide $12x^4y^6$ by $3xy^4$.

SOLUTION

$$\frac{12x^4y^6}{3xy^4} = 4x^{4-1}y^{6-4} = 4x^3y^2$$

Math Note: $x^0 = 1$ $(x \neq 0)$.

EXAMPLE
Divide $18x^3y^2z^4$ by $3xy^2z^3$.

SOLUTION

$$\frac{18x^3y^2z^4}{3xy^2z^3} = 6x^{3-1}y^{2-2}z^{4-3} = 6x^2z$$

PRACTICE
Divide:

1. $36m^3n^4 \div 9m^2n$
2. $-40x^5y^4z^5 \div 8x^3y^2z^2$
3. $15a \div 3$
4. $49a^4b^5 \ (-7ab^4)$
5. $-21c^4d^2 \div (-7cd)$

ANSWERS

1. $4mn^3$
2. $-5x^2y^2z^3$
3. $5a$
4. $-7a^3b$
5. $3c^3d$

Division of a Polynomial by a Monomial

To divide a polynomial by a monomial, divide each term in the polynomial by the monomial.

EXAMPLE
Divide $15x^2 + 10x$ by $5x$.

SOLUTION

$$\frac{15x^2 + 10x}{5x} = \frac{15x^2}{5x} + \frac{10x}{5x} = 3x + 2$$

EXAMPLE
Divide $6x^2y^3 + 12x^4y^2 \div (-3xy)$

SOLUTION

$$\frac{6x^2y^3 + 12x^4y^2}{-3xy} = \frac{6x^2y^3}{-3xy} + \frac{12x^4y^2}{-3xy} = -2xy^2 - 4x^3y$$

EXAMPLE
Divide $15x^3 - 10x^2 + 5x$ by 5.

SOLUTION

$$\frac{15x^3 - 10x^2 + 5x}{5} = \frac{15x^3}{5} + \frac{-10x^2}{5} + \frac{5x}{5} = 3x^3 - 2x^2 + x$$

PRACTICE
Divide:

1. $(21x^2 - 14x) \div 7x$
2. $(30x^2 + 20x + 15) \div 5$
3. $(18x^2y^3 + 12x^4y^2) \div (-6x)$
4. $(a^3b^2 + 2a^4b - 5a^2b^2) \div ab$
5. $(8x^3y^2z - 12x^4yz^2) \div (-4xyz)$

ANSWERS

1. $3x - 2$
2. $6x^2 + 4x + 3$
3. $-3xy^3 - 2x^3y^2$
4. $a^2b + 2a^3 - 5ab$
5. $-2x^2y + 3x^3z$

Quiz

1. Add $(2x^2 + x + 1) + (3x^2 - 5x + 2)$.
   (a) $5x^2 - 4x + 3$
   (b) $6x^2 - 4x + 2$
   (c) $6x^2 - 5x + 2$
   (d) $5x^2 - 6x + 3$

2. Add $(6x + 2y) + (4x - 8y)$.
   (a) $10x + 10y$
   (b) $24x - 32y$
   (c) $10x - 6y$
   (d) $10x - 16y$

3. Subtract $(2x - 8) - (5x + 4)$.
   (a) $10x - 32$
   (b) $-3x - 12$
   (c) $3x + 12$
   (d) $-3x - 4$

4. Subtract $(3a - 2b + c) - (4a + 5b + 6c)$.
   (a) $-12a - 10b + 6c$
   (b) $7a + 3b + 6c$
   (c) $a - 3b - 7c$
   (d) $-a - 7b - 5c$

5. Multiply $7x^2 \cdot 3x$.
   (a) $10x^3$
   (b) $21x^3$
   (c) $21x^2$
   (d) $10x^2$

6. Multiply $3x^2 \cdot 4xy \cdot (-2x^3)$.
   (a) $12\ x^3 y$
   (b) $-24x^6 y$
   (c) $10x^2 y^3$
   (d) $-24x^3 y$

7. Find $(3x^2)^3$.
   (a) $27x^6$
   (b) $9x^6$

(c) $27x^5$

(d) $9x^5$

8. Find $(-2x^2y^3)^4$.
   (a) $-16x^6y^{12}$
   (b) $-8x^6y^7$
   (c) $8x^6y^3$
   (d) $16x^8y^{12}$

9. Multiply $y(2y^2 - y + 6)$.
   (a) $3y^3 - 3y^2 + 6$
   (b) $2y^3 - y^2 + 6y$
   (c) $3y^3 - 4y^2 + 7y$
   (d) $2y^3 + 3y^2 + 6y$

10. Multiply $-5a(6a^2 + 3ab + 2b)$.
   (a) $-11a^3 - 2a^2b - 3ab$
   (b) $11a^3 + 8a^2b + 7ab$
   (c) $-30a^3 - 15a^2b - 10ab$
   (d) $30a^3 + 15a^2b + 10ab$

11. Multiply $(x - 9)(2x - 1)$.
   (a) $2x^2 - 19x + 9$
   (b) $2x^2 + 9$
   (c) $3x^2 - 10$
   (d) $2x^2 + 19x + 9$

12. Multiply $(4x - 7)(2x + 1)$.
   (a) $8x^2 - 14x - 7$
   (b) $6x^2 - 7$
   (c) $8x^2 - 7$
   (d) $8x^2 - 10x - 7$

13. Find $(x + 7)^2$.
   (a) $x^2 + 49$
   (b) $x^2 + 14x + 49$
   (c) $x^2 + 7x + 49$
   (d) $2x^2 + 14$

14. Find $(3x - 8)^2$.

(a) $9x^2 - 24x + 64$

(b) $9x^2 - 64$

(c) $9x^2 - 48x + 64$

(d) $9x^2 + 64$

15. Multiply $(3a^2 + 2a - 1)(5a - 1)$.
   (a) $15a^3 + 13a^2 + 7a - 1$
   (b) $15a^3 + 7a^2 - 10a + 1$
   (c) $15a^3 - 13a^2 - 10a - 1$
   (d) $15a^3 + 7a^2 - 7a + 1$

16. Multiply $(x - 3)(x^2 + 4x - 6)$.
   (a) $x^3 + x^2 - 18x + 18$
   (b) $x^3 - x^2 + 18x - 18$
   (c) $x^3 + x^2 - 18x - 18$
   (d) $x^3 + x^2 + 18x + 18$

17. Divide $15x^3 \div 3x$.
   (a) 5
   (b) $5x^2$
   (c) $5x$
   (d) $5x^3$

18. Divide $-32a^2b^3c^4 \div 8ab^3c^2$.
   (a) $-4ac^2$
   (b) $4ab^2c$
   (c) $-4abc$
   (d) $4a^2bc$

19. Divide $32x^2 + 30x - 6$ by 2.
   (a) $16x^2 - 15x - 3$
   (b) $16x^2 + 15x - 3$
   (c) $16x^2 + 30x - 6$
   (d) $16x^2 - 15x + 6$

20. Divide $25a^2b^3 - 20\,a^4b^2 + 15ab$ by $-5ab$
   (a) $5ab - 4a^2b - 3$
   (b) $-5a^2b + 4ab - 3$
   (c) $5ab^2 + 4a^3b - 3$
   (d) $-5ab^2 + 4a^3b - 3$

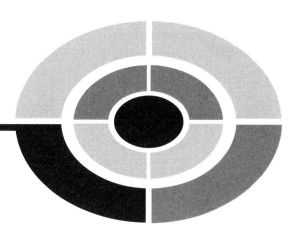

Final Exam

1. Name 5,200,003.
   - (a) fifty-two thousand, three
   - (b) five million, two hundred thousand, three
   - (c) fifty-two million, two hundred three thousand
   - (d) five million, two hundred three thousand

2. Round 63,286 to the nearest thousand.
   - (a) 62,000
   - (b) 63,300
   - (c) 63,000
   - (d) 63,200

3. Add 1,256 + 23,840 + 506.
   - (a) 25,602
   - (b) 35,623
   - (c) 25,503
   - (d) 36,415

4. Subtract 451,203 − 9,177.
   - (a) 432,206

    (b) 431,062

    (c) 442,126

    (d) 442,026

5. Multiply $5{,}126 \times 372$.
- (a) 190,672
- (b) 1,906,272
- (c) 1,906,782
- (d) 1,906,872

6. Divide $133{,}328 \div 641$.
- (a) 208
- (b) 280
- (c) 282
- (d) 228

7. Divide $5321 \div 47$.
- (a) 113 R8
- (b) 123 R10
- (c) 113 R10
- (d) 132 R8

8. Find the total number of calculators in an order if there are 18 boxes with 12 calculators in each box.
- (a) 30
- (b) 216
- (c) 240
- (d) 108

9. Find $|+4|$
- (a) 4
- (b) -4
- (c) 0
- (d) $|-4|$

10. Find the opposite of -5.
- (a) $|5|$
- (b) -5
- (c) 5
- (d) 0

11. Add $-3 + 3$.
- (a) -6
- (b) $+6$

    (c) −9

    (d) 0

12. Subtract $-4 - (-14)$.

    (a) −18

    (b) −10

    (c) 10

    (d) 18

13. Multiply $(-3)(2)(-5)$.

    (a) −30

    (b) 30

    (c) 25

    (d) 18

14. Divide $(-24) \div (3)$.

    (a) 3

    (b) −3

    (c) 8

    (d) −8

15. Simplify $3 \times 4^2 + 6 - 2$.

    (a) 148

    (b) 44

    (c) 52

    (d) 36

16. Find $(-6)^3$.

    (a) −18

    (b) −216

    (c) 216

    (d) 18

17. Reduce $\frac{57}{95}$ to lowest terms.

    (a) $\dfrac{3}{5}$

    (b) $\dfrac{2}{3}$

    (c) $\dfrac{3}{4}$

    (d) $\dfrac{5}{6}$

18. Change $\frac{5}{7}$ to an equivalent fraction in higher terms.

    (a) $\dfrac{35}{42}$

    (b) $\dfrac{16}{21}$

    (c) $\dfrac{15}{28}$

    (d) $\dfrac{35}{49}$

19. Change $\frac{9}{5}$ to a mixed number.

    (a) $1\frac{1}{5}$

    (b) $4\frac{1}{5}$

    (c) $1\frac{4}{5}$

    (d) $1\frac{9}{14}$

20. Change $6\frac{2}{3}$ to an improper fraction.

    (a) $\dfrac{20}{3}$

    (b) $\dfrac{11}{3}$

    (c) $\dfrac{12}{3}$

    (d) $\dfrac{20}{6}$

21. Add $\frac{5}{9} + \frac{7}{12}$.

    (a) $\dfrac{12}{21}$

    (b) $1\frac{5}{36}$

    (c) $\dfrac{35}{108}$

    (d) $1\frac{7}{36}$

22. Subtract $\frac{14}{15} - \frac{3}{20}$.

    (a) $\dfrac{11}{35}$

    (b) $\dfrac{56}{60}$

    (c) $\dfrac{17}{35}$

    (d) $\dfrac{47}{60}$

23. Multiply $\frac{9}{10} \times \frac{5}{9}$.

    (a) $1\frac{31}{50}$

    (b) $\dfrac{1}{2}$

    (c) $\dfrac{50}{81}$

    (d) $\dfrac{2}{3}$

24. Divide $\frac{3}{4} \div \frac{7}{8}$.

    (a) $\dfrac{6}{7}$

    (b) $\dfrac{21}{32}$

    (c) $\dfrac{21}{2}$

    (d) $\dfrac{5}{7}$

25. Add $2\frac{3}{4} + 1\frac{5}{6} + 4\frac{1}{8}$.

    (a) $7\frac{13}{24}$

    (b) $7\frac{3}{8}$

(c) $8\frac{17}{24}$

(d) $9\frac{1}{12}$

26. Subtract $10\frac{1}{3} - 6\frac{4}{5}$.

(a) $3\frac{8}{15}$

(b) $4\frac{7}{15}$

(c) $26\frac{13}{15}$

(d) $3\frac{7}{15}$

27. Multiply $2\frac{1}{2} \cdot 5\frac{7}{8}$.

(a) $10\frac{7}{16}$

(b) $10\frac{1}{8}$

(c) $14\frac{11}{16}$

(d) $11\frac{1}{4}$

28. Divide $15\frac{2}{3} \div 3\frac{1}{3}$.

(a) $52\frac{2}{9}$

(b) $5\frac{1}{3}$

(c) 7

(d) $4\frac{7}{10}$

29. How many pieces of ribbon $1\frac{3}{4}$ inches long can be cut from a piece that is $12\frac{1}{4}$ inches long?

(a) 5

(b) 6

(c) 7

(d) 8

30. Simplify $3 \times 2\frac{1}{8} \div \frac{3}{4} + 5$.

(a) $11\frac{1}{3}$

(b) $13\frac{1}{2}$

(c) $9\frac{7}{8}$

(d) $10\frac{1}{4}$

31. In the number 18.63278, the place value of the 2 is:
    (a) tenths
    (b) hundredths
    (c) thousandths
    (d) ten thousandths

32. Name the number 0.00086.
    (a) eighty-six millionths
    (b) eighty-six ten thousandths
    (c) eighty-six thousandths
    (d) eighty-six hundred thousandths

33. Round 0.23714 to the nearest hundredths.
    (a) 0.237
    (b) 0.24
    (c) 0.23
    (d) 0.2

34. Add 0.25 + 1.263 + 0.0348.
    (a) 1.5478
    (b) 2.3624
    (c) 0.9713
    (d) 1.6327

35. Subtract 0.531 − 0.0237.
    (a) 0.6341
    (b) 0.5073
    (c) 0.0514
    (d) 0.6032

36. Multiply 0.35 × 0.006.
    (a) 0.021
    (b) 0.0021
    (c) 0.21
    (d) 0.00021

37. Divide 2.0475 ÷ 0.325.
    (a) 63
    (b) 6.3
    (c) 0.63
    (d) 0.063

38. Arrange in order from smallest to largest: 0.416, 0.061, 1.64, 0.003.
    (a) 0.416, 1.64, 0.003, 0.061

(b) 1.64, 0.416, 0.061, 0.003
(c) 0.003, 0.416, 0.416, 1.64
(d) 0.003, 0.061, 0.416, 1.64

39. Change $\frac{19}{22}$ to a decimal.

(a) $0.\overline{863}$
(b) 0.863
(c) $0.8\overline{63}$
(d) $0.8\overline{36}$

40. Change 0.64 to a fraction in lowest terms.

(a) $\dfrac{3}{5}$

(b) $\dfrac{16}{25}$

(c) $\dfrac{5}{6}$

(d) $\dfrac{64}{100}$

41. Multiply $\frac{1}{8} \times 0.62$.

(a) 0.275

(b) $\dfrac{7}{8}$

(c) 0.0775

(d) $\dfrac{31}{40}$

42. If a calculator costs $9.98 and a notebook costs $1.89, find the cost of 2 calculators and 3 notebooks.
(a) $25.63
(b) $11.87
(c) $35.61
(d) $19.43

43. Write 0.006 as a percent.
(a) 6%
(b) 60%
(c) 0.6%

(d) 0.006%

44. Write $\frac{14}{25}$ as a percent.

(a) 56%
(b) 0.56%
(c) 5.6%
(d) 560%

45. Write 32% as a fraction.

(a) $\frac{3}{5}$

(b) $\frac{8}{15}$

(c) $\frac{5}{16}$

(d) $\frac{8}{25}$

46. Find 16% of 64.
(a) 10.24
(b) 40
(c) 400
(d) 0.25

47. 40% of what number is 84?
(a) 33.6
(b) 48
(c) 50.4
(d) 210

48. 18 is what percent of 24?
(a) $133.\overline{3}\%$
(b) 75%
(c) 25%
(d) 125%

49. The finance rate charged on a credit card is 1.5%. Find the finance
     charge if the balance is $150.00.
(a) $22.50
(b) $225
(c) $2.25

(d) $10.00

50. If the sales tax on $320.00 is $20.80, find the sales tax rate.
   (a) 5%
   (b) 6%
   (c) 6.5%
   (d) 7%

51. Evaluate $-3x + 2xy$ when $x = -5$ and $y = 6$.
   (a) -25
   (b) 35
   (c) -45
   (d) 65

52. Multiply $-6(2x - 8y + 10)$.
   (a) $-12x + 48y - 60$
   (b) $-12x - 48y + 60$
   (c) $12x - 48y + 60$
   (d) $12x + 48y - 60$

53. Combine like terms $-5a + 2b - 7c + 6c - 3b + 3a$.
   (a) $8a + 5b - 10c$
   (b) $-2a - b - c$
   (c) $-8a - 5b + 10c$
   (d) $2a + b - c$

54. Combine like terms $7(3x - 8) + 2(x - 5)$.
   (a) $31x - 56$
   (b) $-23x + 56$
   (c) $31x - 66$
   (d) $23x - 66$

55. Find the Celsius temperature (°C) when the Fahrenheit temperature is 86°. Use $C = \frac{5}{9}(F - 32)$.
   (a) $42°$
   (b) $30°$
   (c) $15°$
   (d) $12°$

56. Solve $3x - 8 = 31$.
   (a) 13
   (b) 7
   (c) 15
   (d) 9

57. Solve $-2(3x - 6) + 18 = 24$.
    (a) 6
    (b) −4
    (c) 2
    (d) 1

58. If the sum of 3 times a number and 2 times a number is 60, find the number.
    (a) 8
    (b) 10
    (c) 2
    (d) 12

59. The ratio of 20 to 8 is:
    (a) $\dfrac{3}{5}$

    (b) $\dfrac{5}{2}$

    (c) $\dfrac{2}{5}$

    (d) $\dfrac{3}{2}$

60. Find the value of x when $\dfrac{x}{6} = \dfrac{12}{24}$.
    (a) 18
    (b) 3
    (c) 16
    (d) 24

61. If a person uses 15 gallons of gasoline to travel 345 miles, how many gallons of gasoline will be needed to travel 506 miles?
    (a) 28 gallons
    (b) 22 gallons
    (c) 27 gallons
    (d) 34 gallons

62. Find the circumference of a circle whose diameter is 28 inches. Use $\pi = 3.14$.
    (a) 87.92 in.
    (b) 21.98 in.
    (c) 15.62 in.

(d) 43.96 in.

63. Find the perimeter of a triangle whose sides are 8 in., 9 in., and 10 in.
     (a) 13.5 in.
     (b) 19 in.
     (c) 27 in.
     (d) 17 in.

64. Find the perimeter of a square whose side is 7.5 inches.
     (a) 30 in.
     (b) 15 in.
     (c) 22.5 in.
     (d) 56.25 in.

65. Find the perimeter of a rectangle whose length is 18 feet and whose width is 9.6 feet.
     (a) 55.2 ft
     (b) 172.8 ft
     (c) 86.4 ft
     (d) 27.6 ft

66. Find the area of a circle whose radius is 19 yards. Use $\pi = 3.14$.
     (a) 59.66 yd^2
     (b) 4534.16 yd^2
     (c) 1133.54 yd^2
     (d) 119.32 yd^2

67. Find the area of a square whose side is $9\frac{1}{4}$ inches.
     (a) 37 in.^{2}
     (b) 18.5 in.^{2}
     (c) 42.6 in.^{2}
     (d) $85\frac{9}{16}$ in.^{2}

68. Find the area of a trapezoid whose height is 14.3 inches and whose bases are 9 inches and 12 inches.
     (a) 1544.4 in.^{2}
     (b) 150.15 in.^{2}
     (c) 300.3 in.^{2}
     (d) 450.2 in.^{2}

69. Find the volume of a sphere whose radius is 27 inches. Use $\pi = 3.14$.
     (a) 82,406.16 in.^{3}
     (b) 84.78 in.^{3}
     (c) 28.26 in.^{3}

(d) 798.62 in.^{3}

70. Find the volume of a cylinder whose height is 12 feet and whose radius is 3 feet. Use $\pi = 3.14$.
(a) 28.26 ft^3
(b) 339.12 ft^3
(c) 37.68 ft^3
(d) 1356.48 ft^3

71. Find the volume of a pyramid whose base is 14 inches by 12 inches and whose height is 10 inches.
(a) 560 in.^{3}
(b) 1820 in.^{3}
(c) 1400 in.^{3}
(d) 910 in.^{3}

72. Find the length of the hypotenuse of a right triangle if its sides are 5 yards and 12 yards.
(a) 15 in.
(b) 16 in.
(c) 14 in.
(d) 13 in.

73. How many square inches are in 15 square feet?
(a) 180 in.^{2}
(b) 720 in.^{2}
(c) 2160 in.^{2}
(d) 1440 in.^{2}

74. Change 15 yards to inches.
(a) 180 in.
(b) 540 in.
(c) 2160 in.
(d) 45 in.

75. Change 84 feet to yards.
(a) 28 yd
(b) 252 yd
(c) 7 yd
(d) 14 yd

76. Change 7 feet 8 inches to inches.
(a) 84 in.
(b) 80 in.

(c) 92 in.

(d) 96 in.

77. Change 150 pounds to ounces.
    (a) 450 oz
    (b) 2400 oz
    (c) 4800 oz
    (e) 900 oz

78. Change 1248 ounces to pounds.
    (a) 39 lb
    (b) 78 lb
    (c) 16 lb
    (d) 117 lb

79. Change 4.5 tons to ounces.
    (a) 9000 oz
    (b) 144,000 oz
    (c) 720 oz
    (d) 72 oz

80. Change 6.4 quarts to pints.
    (a) 12.8 pt
    (b) 25.6 pt
    (c) 3.2 pt
    (d) 19.2 pt

81. Change 78 months to years.
    (a) 936 years
    (b) 468 years
    (c) 19.5 years
    (d) 6.5 years

82. Change 7 miles to feet.
    (a) 12,320 ft
    (b) 18,489 ft
    (c) 36,960 ft
    (d) 6160 ft

83. In which quadrant is the point $(-6, 3)$ located?
    (a) Q I
    (b) Q II
    (c) Q III
    (d) Q IV

84. Which equation represents a vertical line?
    (a) $x = -5$
    (b) $4x - 7y = 21$
    (c) $y = 8$
    (d) $2x + y = 6$

85. Find y when $x = -3$ for $6x + 2y = -10$.
    (a) -5
    (b) 4
    (c) 5
    (d) -4

86. Which is a solution to $9x - 2y = 18$?
    (a) $(-4, -9)$
    (b) $(-2, 0)$
    (c) $(0, 9)$
    (d) $(4, 9)$

87. Find the slope of the line containing two points whose coordinates are $(-4, -3)$ and $(8, 2)$.

    (a) $-\dfrac{5}{12}$

    (b) $\dfrac{5}{12}$

    (c) $-\dfrac{12}{5}$

    (d) $\dfrac{12}{5}$

88. Find the slope of a line whose equation is $5x + 8y = 29$.

    (a) $-\dfrac{5}{8}$

    (b) $\dfrac{8}{5}$

    (c) $-\dfrac{8}{5}$

    (d) $\dfrac{5}{8}$

89. The slope of a horizontal line is:
    (a) 0
    (b) 1
    (c) −1
    (d) undefined

90. Find the y intercept of the line $3x - 8y = 24$.
    (a) $(0, -3)$
    (b) $(8, 0)$
    (c) $(0, 3)$
    (d) $(-8, 0)$

91. Add $(8x^2 + 3x - 2) + (9x^2 - 7x + 8)$.
    (a) $17x^2 + 4x - 6$
    (b) $17x^2 - 10x + 6$
    (c) $17x^2 - 4x + 10$
    (d) $17x^2 - 4x + 6$

92. Subtract $(3x - 2y - 10) - (5x - 10y - 15)$.
    (a) $-2x + 8y + 5$
    (b) $8x - 12y + 25$
    (c) $-2x + 8y - 5$
    (d) $-2x - 8y + 5$

93. Multiply $9x \cdot 7x^3 \, y \cdot 2y$.
    (a) $63x^3y^2$
    (b) $126x^4y^2$
    (c) $126x^3y$
    (d) $63x^4y^2$

94. Find $(-3x^3y^4)^2$.
    (a) $-9x^6y^8$
    (b) $-3x^6y^8$
    (c) $9x^6y^8$
    (d) $9x^9y^{16}$

95. Multiply $2x(3x - 5y + 6)$.
    (a) $6x - 10y + 12$
    (b) $6x - 10xy + 12$
    (c) $6x^2 + 10xy - 12$
    (d) $6x^2 - 10xy + 12x$

96. Multiply $(9x - 3)(7x + 5)$.
    (a) $63x^2 + 24x - 15$
    (b) $63x^2 - 15$
    (c) $63x^2 - 24x - 14$
    (d) $63x^2 - 24x + 15$

97. Find $(2x - 4)^2$.
    (a) $4x - 16$
    (b) $4x^2 - 16x + 16$
    (c) $4x^2 - 8x + 16$
    (d) $4x^2 - 6x + 16$

98. Multiply $(4x^2 - 2x + 6)(x - 1)$.
    (a) $4x^3 - 2x^2 + 8x - 6$
    (b) $4x^3 - 6x^2 - 8x + 6$
    (c) $4x^3 + 2x^2 + 4x - 6$
    (d) $4x^3 - 6x^2 + 8x - 6$

99. Divide $48x^3y^2z^5 \div 16xyz^2$.
    (a) $3x^4y^6z^7$
    (b) $3xyz$
    (c) $3x^2yz^3$
    (d) $3x^3y^2z^4$

100. Divide $15a^2b^2c^2 - 10\,abc^3$ by $-5ac$.
    (a) $3ab^2c + 2bc^2$
    (b) $-3ac + 2bc$
    (c) $3abc - 2bc^2$
    (d) $-3ab^2c + 2bc^2$

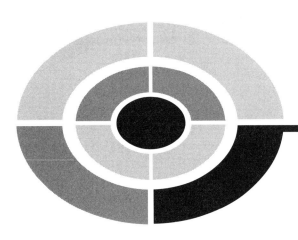

Answers to Chapter Quizzes

CHAPTER 1

1. c	2. d	3. a	4. d	5. b
6. b	7. d	8. a	9. c	10. a
11. b	12. c	13. d	14. d	15. c
16. a	17. d	18. b	19. c	20. a

CHAPTER 2

1. a	2. c	3. c	4. d	5. d
6. d	7. a	8. b	9. a	10. b
11. c	12. a	13. d	14. b	15. b
16. d	17. c	18. b	19. d	20. a
21. d				

CHAPTER 3

1. c   2. b   3. b   4. d   5. b
6. d   7. a   8. b   9. d   10. b
11. a 12. c 13. b 14. a 15. c
16. d 17. a 18. d 19. a 20. b

CHAPTER 4

1. c   2. a   3. b   4. d   5. b
6. a   7. d   8. d   9. b   10. c
11. d 12. a 13. d 14. b 15. c
16. c 17. c 18. a 19. d 20. a

CHAPTER 5

1. c   2. c   3. b   4. d   5. b
6. a   7. a   8. c   9. a   10. d
11. c 12. a 13. c 14. c 15. c
16. b 17. a 18. a 19. b 20. b

CHAPTER 6

1. b   2. a   3. b   4. b   5. c
6. c   7. c   8. d   9. c   10. d
11. a 12. b 13. c 14. b 15. d
16. c 17. b 18. b 19. b 20. d

CHAPTER 7

1. b   2. d   3. a   4. c   5. c
6. d   7. b   8. c   9. b   10. a
11. c 12. c 13. b 14. a 15. d
16. b 17. d 18. c 19. d 20. a

CHAPTER 8

1. c   2. a   3. c   4. d   5. d
6. d   7. a   8. b   9. b   10. c
11. c 12. b 13. c 14. a 15. c
16. d 17. b 18. a 19. b 20. c

CHAPTER 9

1. c   2. b   3. d   4. a   5. a
6. b   7. a   8. c   9. a   10. c
11. b   12. a   13. b   14. b   15. d
16. a   17. c   18. b   19. d   20. b

CHAPTER 10

1. c   2. b   3. a   4. d   5. c
6. a   7. d   8. b   9. d   10. b
11. d   12. d   13. c   14. b   15. d
16. c   17. a   18. c   19. c   20. c

CHAPTER 11

1. c   2. a   3. b   4. b   5. d
6. c   7. b   8. a   9. d   10. c
11. c   12. c   13. d   14. a   15. b
16. b   17. c   18. c   19. d   20. c

CHAPTER 12

1. a   2. c   3. b   4. d   5. b
6. b   7. a   8. d   9. b   10. c
11. a   12. d   13. b   14. c   15. d
16. a   17. b   18. a   19. b   20. d

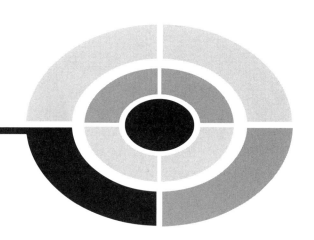

Final Exam Answers

1. b	2. c	3. a	4. d	5. d
6. a	7. c	8. b	9. a	10. c
11. d	12. c	13. b	14. d	15. c
16. b	17. a	18. d	19. c	20. a
21. b	22. d	23. b	24. a	25. c
26. a	27. c	28. d	29. c	30. b
31. c	32. d	33. b	34. a	35. b
36. b	37. b	38. d	39. c	40. b
41. c	42. a	43. c	44. a	45. d
46. a	47. d	48. b	49. c	50. c
51. c	52. a	53. b	54. d	55. b
56. a	57. d	58. d	59. b	60. b
61. b	62. a	63. c	64. a	65. a
66. c	67. d	68. b	69. a	70. b
71. a	72. d	73. c	74. b	75. a
76. c	77. b	78. b	79. b	80. a
81. d	82. c	83. b	84. a	85. b
86. d	87. b	88. a	89. a	90. a
91. d	92. a	93. b	94. c	95. d
96. a	97. b	98. d	99. c	100. d

Overcoming Math Anxiety

Welcome to the program on overcoming math anxiety. This program will help you to become a successful student in mathematics. It is divided into three parts:

- Part I will explain the nature and causes of math anxiety.
- Part II will help you overcome any anxiety that you may have about mathematics.
- Part III will show you how to study mathematics and how to prepare for math exams.

As you will see in this program, the study skills that you need to be successful in mathematics are quite different from the study skills that you use for other classes. In addition, you will learn how to use the classroom, the textbook, and yes, even the teacher, to increase your chances of success in mathematics.

Part I   The Nature and Causes of Math Anxiety

Many students suffer from what is called "math anxiety." Math anxiety is very real, and it can hinder your progress in learning mathematics. Some of the physical symptoms of math anxiety include:

- nervousness;
- pounding heart;
- rapid breathing;
- sweating;
- nauseousness;
- upset stomach; and
- tenseness.

In addition to the physical symptoms, people may experience any or all of the following mental symptoms:

- a feeling of panic or fear;
- cloudy or fuzzy thinking;
- lack of concentration;
- a mental block in thinking; and
- feelings of helplessness, guilt, shame, inferiority, or stupidity.

If you have any of these physical or mental symptoms when you are in a mathematics classroom, during a mathematics test, or when you are doing your mathematics homework, then you suffer some degree of math anxiety.

Naturally, a little fear or uneasiness accompanies all of us when we take exams, but if these symptoms are severe enough to keep you from doing your best, it is time to do something about your math anxiety.

In order to decrease your anxiety, it is helpful to determine why you have math anxiety. After all, no student, as far as we know, is born with math anxiety. Let's look at some possible causes of math anxiety, and while you are reading them, think about some reasons that might have caused you to develop math anxiety.

REASON 1: POOR MATH TEACHERS

Throughout elementary school, middle school, and even in high school, you have had many mathematics teachers. Some were good, and some were not. Sometimes teachers are forced to teach math when they are not trained for it or when they dislike the subject themselves, or even when they do not understand it. Naturally, in these situations, teachers cannot do a good job. Think about your teachers. Were they good or were they not?

REASON 2: TRAUMATIC EXPERIENCES

Many students can recall an incident in their education involving mathematics where they had a traumatic experience. Were you ever called to the blackboard to do a math problem in front of the whole classroom and were unable to do it? Did your teacher belittle you for your inability in mathematics? Were you ever punished for not doing your mathematics homework?

Sometimes teachers, parents, or tutors can make you feel stupid when you ask questions. One such traumatic experience in a person's life can cause that person to fear or hate mathematics for the rest of his or her life.

One student recalled that when he was studying fractions in third grade, the teacher picked him up by his ankles and turned him upside down to illustrate the concept of inverting fractions!

Can you recall a traumatic event that happened to you that was related to mathematics?

REASON 3: PROLONGED ABSENCE FROM SCHOOL

Students who miss a lot of school usually fall behind in mathematics. Many times, they fall so far behind they cannot catch up and end up failing the course. This is because mathematics is a cumulative subject. What you learn today you will use tomorrow. If you fail to learn a topic because you were absent, when the time comes to use the material to learn something new, you will be unable to do so. It's like trying to build a second storey on a building with a weak or inadequate first storey. It cannot be done. For example, if you look at long division, it involves multiplication and subtraction. So if you can't multiply or subtract, you will not be able to do long division. All mathematics is like this!

Think about a time when you were absent from school. Were you lost in math class when you came back?

REASON 4: POOR SELF-IMAGE

Occasionally, students with a poor self-image have difficulty with mathematics. This poor self-image could have been acquired when students were told something like this:

- "Men can do math better than women."
- "Math requires logic, and you're not very logical."
- "You should be able to do math in your head."
- "Boy, that's a stupid question."

These comments can be made by parents, teachers, spouses, and even friends. They can make a person feel very inadequate when doing mathematics.

I have heard that it takes ten positive comments to overcome one negative comment.

Have you ever been made to feel stupid by someone's negative comment?

REASON 5: EMPHASIS ON THE CORRECT ANSWER

Mathematics involves solving problems, and solving problems involves getting the correct answers. When doing mathematics, it is very easy to make a simple mistake. This, in turn, will lead to the wrong answer. Even when you use a calculator, it is very easy to press the wrong key and get an incorrect answer.

Furthermore, when students are under pressure in a testing situation, they become nervous and tend to make more mistakes than they would on their homework exercises. When this occurs, students may feel that it is impossible to learn mathematics. This can cause a great deal of anxiety.

Have you ever failed a test, even when you knew how to do the problems but made simple mistakes?

REASON 6: PLACEMENT IN THE WRONG COURSE

In college, it is easy to sign up for the wrong mathematics course – i.e., for a course in which you do not have enough math skills to succeed. Therefore, it is necessary to be sure that you are properly prepared for the course that you are going to take.

Most mathematics courses, except the introductory ones, have **prerequisites**. For example, in order to succeed in algebra, you need to have an understanding of arithmetic. Success in trigonometry requires knowledge of algebra and geometry.

Before signing up for a mathematics course, check the college catalog to see if the course has a prerequisite. If so, make sure that you have successfully completed it. **Do not skip courses**. If you do, you will probably fail the course.

It would be like trying to take French III without taking French I and French II.

If you are enrolled in a mathematics course now, have you completed the prerequisite courses? If you are planning to take a mathematics course, did you check the college catalog to see if you have completed the prerequisite courses?

REASON 7: THE NATURE OF MATHEMATICS

Mathematics is unlike any other course in that it requires more skills than just memorization. Mathematics requires you to use analytical reasoning skills, problem solving skills, and critical thinking skills. It is also abstract in nature since it uses symbols. In other words, you have to do much more than just memorize a bunch of rules and formulas to be successful in mathematics.

Many students view mathematics with the attitude of, "Tell me what I have to memorize in order to pass the test." If this is the way you think about mathematics, you are on the road to failure. It is time to change your way of thinking about mathematics. From now on, realize that you have to do far more than just memorize.

How do you view mathematics?

Of course, there are many other factors that can cause math anxiety and lead to poor performance in mathematics. Each person has his or her own reasons for math anxiety. Perhaps you would like to add your own personal reasons for your math anxiety.

Part II   Overcoming Math Anxiety

Now that we have explained the nature of math anxiety and some of the possible causes of it, let us move to the next part of this program: How You Can Overcome Math Anxiety. The plan for overcoming math anxiety consists of two parts. **First**, you need to develop a **positive attitude** towards mathematics; **second**, you need to learn some **calming** and **relaxation skills** to use in the classroom, when you are doing your homework, and especially when you are taking a math test.

DEVELOPING A POSITIVE ATTITUDE

To be successful in mathematics, you need to develop a positive attitude towards the subject. Now you may ask, "How can I develop a positive attitude for something I hate?"

Here's the answer.

First, it is important to **believe in yourself**. That is, believe that you have the ability to succeed in mathematics. Dr. Norman Vincent Peale wrote a book entitled *You Can If You Think You Can*. In it he states, "Remember that self-trust is the first secret of success. So trust yourself." This means that in order to succeed, you must believe absolutely that you can.

"Well," you say, "how can I start believing in myself?"

This is possible by using affirmations and visualizations.

An **affirmation** is a positive statement that we make about ourselves. Remember the children's story entitled *The Little Engine That Could*? In the story, the little engine keeps saying, "I think I can, I think I can" when it was trying to pull the cars up the mountain. Although this is a children's story, it illustrates the use of affirmations. Many athletes use affirmations to enhance their performance. You can, too. Here are a few affirmations you can use for mathematics:

- "I believe I can succeed in mathematics."
- "Each day I am learning more and more."
- "I will succeed in this course."
- "There is nothing that can stop me from learning mathematics."

In addition, you may like to write some of your own. I think you can see what I mean. Here are some suggestions for writing affirmations:

1. State affirmations in the positive rather than the negative. For example, don't say, "I will not fail this course." Instead, affirm, "I will pass this course."
2. Keep affirmations as short as possible.
3. When possible, use your name in the affirmation. Say, "I (your name) will pass this course."
4. Be sure to write down your affirmations and read them every morning when you awake and every evening before going to sleep.

Affirmations can also be used before entering math class, before doing your homework, and before the exam. You can use affirmations to relax when your symptoms of math anxiety start to occur. This will be explained later.

Another technique that you can use to develop a positive attitude and help you in believing in yourself is to use **visualizations**. When you visualize, you create a mental picture of yourself succeeding in your endeavor. Whenever you visualize, try to use as many of your senses as you can – sight, hearing, feeling, smelling, and touch. For example, you may want to visualize yourself getting an A on the next math test. Paint a picture of a scene like this:

> Visualize yourself sitting in the classroom while your instructor is passing back the test paper. As he or she gives you your test paper, feel the paper in your hand and see the big red A on top of it. Your friend next to you says, "Way to go!" I don't know how to include your sense of smell in this picture, but maybe you can think of a way.

The best time to visualize is after relaxing yourself and when it is quiet. Of course, you can use visualization any time when it is convenient.

A fourth way to develop a positive attitude is to give yourself a **pep talk** every once in a while. Pep talks are especially helpful when you are trying to learn different topics, when you are ready to give up, or when you have a mental block.

How we talk to ourselves is very important in what we believe about ourselves. Shad Helmstetter wrote an entire book on what people say to themselves. It is entitled *The Self-Talk Solution*. In it he states, "Because 75% or more of our early programming was of the negative kind, we automatically followed suit with self-programming of the same negative kind." As you can see, negative self-talk will lead to a negative attitude.

Conversely, positive self-talk will help you develop a positive self-image. This will help you not only in mathematics but also in all aspects of your life.

Another suggestion to help you to be positive about mathematics is to be **realistic**. You should not expect to have perfect test papers every time you take a test. Everybody makes mistakes. If you made a careless mistake, forget it. If you made an error on a process, make sure that you learn the correct procedure before the next test. Also, don't expect to make an A on every exam: sometimes a C is the best that you can accomplish. In mathematics, some topics are more difficult than others, so if you get a C or even a lower grade on one test, resolve to study harder for the next test so that you can bring up your average.

Finally, it is important to develop **enthusiasm for learning**. Learning is to your mind what exercise is to your body. The more you learn, the more intelligent you will be. Don't look at learning as something to dread. When you learn something new and different, you are improving yourself and exercising your mind. Learning can be fun and enjoyable, but it requires effort on your part.

Developing a positive attitude about mathematics alone will not enable you to pass the course. Achieving success requires hard work by studying, being persistent, and being patient with yourself. Remember the old saying, "Rome wasn't built in a day."

LEARNING CALMING AND RELAXATION SKILLS

The second part of this program is designed to show you some techniques for stress management. These techniques will help you to relax and overcome some of the physical and mental symptoms (nervousness, upset stomach, etc.) of math anxiety. The first technique is called **deep breathing**. Here you sit comfortably in a chair with your back straight, feet on the floor, and hands on your knees. Take a deep, slow breath in through your nose and draw it down into your stomach. Feel your stomach expand. Then exhale through your mouth. Take several deep breaths. Note: If you become dizzy, stop immediately.

Another technique that can be used to calm yourself is to use a **relaxation word**. Sit comfortably, clear your mind, and breathe normally. Concentrate on your breathing, and each time you exhale, say a word such as "relax," "peace," "one," or "calm." Select a word that is pleasing and calming to you.

You can also use a technique called **thought stopping**. Any time you start thinking anxious thoughts about mathematics, say to yourself, "STOP," and then try to think of something else. A related technique is called **thought switching**. Here you make up your mind to switch your thoughts to something pleasant instead of thinking anxious thoughts about mathematics.

In the preceding section, I explained the techniques of **visualization** and **affirmations**. You can also use visualizations and affirmations to help you reduce your anxiety. Whenever you become anxious about mathematics, visualize yourself as being calm and collected. If you cannot do this, then visualize a quiet, peaceful scene such as a beautiful, calm lake in the mountains. You can also repeat affirmations such as "I am calm."

The last technique that you can use to relax is called **grounding**. In order to ground yourself, sit comfortably in a chair with your back straight, feet on the floor, and hands on your legs or on the side of the chair. Take several deep breaths and tell yourself to relax. Next, think of each part of your body being grounded. In other words, say to yourself, "I feel my feet grounded to the floor." Then think of your feet touching the floor until you can feel them touching. Then proceed to your ankles, legs, etc., until you reach the top of your head. You can take a deep breath between grounding the various parts of the body, and you can even tell yourself to relax.

In order to learn these techniques, you must practice them over and over until they begin to work.

However, for some people, none of these techniques will work. What then? If you cannot stop your anxious thoughts, then there is only one thing left to do. It is what I call the **brute force technique**. That is to realize that you must attend class, do the homework, study, and take the exams. So force yourself mentally and physically to do what is necessary to succeed. After brute forcing several times, you will see that it will take less effort each time, and soon your anxiety will lessen.

In summary, confront your anxiety head on. Develop a positive attitude and use stress reduction techniques when needed.

Part 3　Study and Test-Taking Techniques

Having an understanding of math anxiety and being able to reduce stress are not enough to be successful in mathematics. You need to learn the basic skills of how to study mathematics. These skills include how to use the classroom, how to use the textbook, how to do your homework, how to review for exams, and how to take an exam.

THE CLASSROOM

In order to learn mathematics, it is absolutely necessary to **attend class**. You should never miss class unless extreme circumstances require it. If you know ahead of time that you will be absent, ask your instructor for the assignment in advance, then read the book, and try to do the homework before the next class. If you have an emergency or are ill and miss class, call your instructor or a friend in the class to get the assignment. Again, read the material and try the problems before the next class, if possible. Be sure to tell your instructor why you were absent.

Finally, if you are going to be absent for an extended period of time, let your instructor know why and get the assignments. If you cannot call him or her, have a friend or parent do it for you.

Come to each class prepared. This means to have all the necessary materials, including homework, notebook, textbook, calculators, pencil, computer disk, and any other supplies you may need.

Always select a seat in front of the classroom and near the center. This assures that you will be able to see the board and hear the instructor.

Pay attention at all times and take good notes. Write down anything your instructor writes on the board. If necessary, bring a tape recorder to class and record the sessions. Be sure to ask your instructor for permission first, though.

You can also ask you instructor to repeat what he or she has said or to slow down if he or she is going too fast. But remember, don't become a pest. You must be reasonable.

Be sure to ask intelligent questions when you don't understand something. Now I know some of you are thinking, "How can I ask intelligent questions when I don't understand?" or, "I don't want to sound stupid."

If you really understand the previous material and you are paying attention, then you can ask intelligent questions.

You must also remember that many times the instructor will leave out steps in solving problems. This is not to make it difficult for you, but you should be able to fill them in. You must be alert, active, and pay complete attention to what's going on in class.

Another important aspect you should realize is that in a class with ten or more students, you cannot expect private tutoring. You cannot expect to understand everything that is taught. But what you must do is copy down everything you can. Later, when you get home, apply the information presented in the chapters of this book.

Be alert, active, and knowledgeable in class.

THE TEXTBOOK

The textbook is an important tool in learning, and you must know how to use it.

It is important to **study** your book. Note that I did not say "read" your book, but I purposely said **study**.

How do you study a mathematics book? It is different from studying a psychology book. First, look at the chapter title. It will tell you what you will be studying in the chapter. For example, if Chapter 5 is entitled "Solving Equations," this means that you will be doing something (solving) to what are called "equations." Next, read the chapter's introduction: it will tell you what topics are contained in the chapter.

Now look at the section headings. Let's say Section 5.3 is entitled "Solving Equations by Using the Multiplication Principle." This tells you that you will be using multiplication to solve a certain type of equation.

Take a pencil and paper and underline in the book all definitions, symbols, and rules. Also, write them down in your notebook.

Now, actually work out each problem that is worked out in the textbook. Do not just copy the problems, but actually try to solve them, following the author's solutions. Fill in any steps the author may have left out. If you do not understand why the author did something, write a note in the book and ask your instructor or a friend to explain it to you. Also, notice how each problem is different from the previous one and what techniques are needed to find the answer. After you have finished this, write the same problems on a separate sheet of paper and try to solve them without looking in your book. Check the results against the author's solutions.

Don't be discouraged if you cannot understand something the first time you read it. Read the selection at least three times. Also, look at your classroom notes. You may find that your instructor has explained the material better than your book. If you still cannot understand the material, do not say, "This book is bad. I can't learn it." What you can do is go to the library and get another book and look up the topic in the table of contents or appendix. Study this author's approach and try to do the problems again.

There is no excuse. If the book is bad, get another one.

Remember that I didn't ever say that learning mathematics was easy. It is not, but it can be done if you put forth the effort!

After you have studied your notes and read the material in the textbook, try to do the homework exercises.

HOMEWORK

Probably the single most important factor which determines success in mathematics is doing the homework. There's an old saying that "Mathematics is not a spectator sport." What this means is that in order to learn mathematics, you must do the homework. As stated previously, it is like learning to play an instrument. If you went to music class but never practiced, you could never learn how to play your instrument. Also, you must practice regularly or you will forget or be unable to play your instrument. Likewise, with mathematics, you must do the homework every day it is assigned. Here are my suggestions for doing your homework:

- First and most important: **DO YOUR HOMEWORK AS SOON AS POSSIBLE AFTER CLASS**. The reason is that the material will still be fresh in your mind.
- Make a habit of studying your mathematics regularly – say, three times a week, five times a week, etc.
- Get your book, notes, and all previous homework problems, calculator, pencil, paper (everything you will need) before you start.
- Do not dally around. Get started at once and do not let yourself be interrupted after you start.

- Concentrate on mathematics only!
- Write the assignment at the top of your paper.
- Read the directions for the problems carefully.
- Copy each problem on your homework paper and number it.
- Do not use scratch paper. Show all of your work on your homework paper.
- Write neatly and large enough. Don't do sloppy work.
- Check your answers with the ones in the back of the book. If no answers are provided, check your work itself.
- See if your answers sound reasonable.
- Write out any questions you have about the homework problems and ask your instructor or another student when you can.
- Draw pictures when possible. This is especially important in courses such as geometry and trigonometry.
- If you have made a mistake, try to locate it. Do not depend on the teacher to find all your mistakes. Make sure that you have copied the problem correctly.
- Don't give up. Doing a problem wrong is better than not doing it at all. (Note: don't spend an exorbitant amount of time on any one problem though.)
- Use any of the special study aids such as summaries, lists of formulas, and symbols that you have made.
- Don't skip steps.
- Finally, if you cannot get the correct answer to a problem, don't stop. Try the next step.

REVIEW

In order to learn mathematics, it is necessary to review before the tests. It is very important to realize one fact. You cannot cram in mathematics. You cannot let your studying go until the night before the exam. If you do, forget it. I have had students who have told me that they spent 3 hours studying before the exam, and then failed it. If those were the only 3 hours they spent studying, there is no way they could learn the material.

Some teachers provide written reviews. Make sure you do them. If this is the case, you can use the review as a practice test. If not, you can make up your own review. Many books have practice tests at the beginning and the end of chapters. If so, you can use these exams as reviews. Some books have extra problems at the end of each chapter. By doing these problems, you have another way to review.

Finally, if there is no review in the book, you can make up your own review by selecting one or two problems from each section in the chapter. Use these problems to make up your own practice test. Be sure not to select the first or second problem in each unit because most mathematics books are arranged so that the easy problems are first.

When you review, it is important to memorize symbols, rules, procedures, definitions, and formulas. In order to memorize, it is best to make a set of cards as shown here:

Front	*Back*
Commutative Property of Addition	For any real numbers, a and b, $a + b = b + a$

On the front of the card, write the name of the property, and on the back, write the property. Then when you are studying, run the cards through the front side and then on the other side. This way you can learn both the property and also the name of the property.

Finally, you must be aware that a review session is not a study session. If you have been doing your work all along, then your review should be short.

TEST-TAKING TIPS

There are three types of exams that mathematics instructors give. They are closed book exams, open book exams, and take home exams.

First, let's talk about closed book exams given in the classroom. Make sure that you show up 5–10 minutes before the class begins. Bring all the necessary materials such as a pencil with an eraser, your calculator, paper, and anything else that you may need. Look over and study materials before class. After you receive your exam paper, look over the entire test before getting started. Read the directions carefully. Do all of the problems that you know how to do first, and then go back and try the others. Don't spend too much time on any one problem. Write down any formulas that you may need. Show all necessary steps if a problem requires it.

If there is any time left after you finish the exam, go back and check it for mistakes.

If you don't understand something, ask your instructor. Remember you cannot expect your instructor to tell you whether or not your answer is correct or how to do the problem.

In open book exams, you should remember that the book is your tool. Do not plan to study the book while you are taking the test. Study it before class. Make sure you know where all the tables, formulas, and rules are in the book. If you are permitted, have the formulas, rules, and summaries written down. Also, have sample problems and their solutions written down.

Don't be elated when your instructor gives you a take home exam. These exams are usually the hardest. You may have to go to the library and get other books on the subject to help you do the problems, or you may have to ask other students for help. The important thing is to get the correct solution. Show all of your work.

When your instructor returns the exam, make a note of the types of problems you missed and go back and review them when you get home.

See your instructor for anything that you are not sure of.

A Final Note

Parts I and II explained the nature and causes of math anxiety and showed some ways to reduce or eliminate the symptoms of math anxiety. Part III explained study, review, and test-taking skills needed for success in mathematics. But there is still one problem left to discuss: What happens if you are in over your head – i.e., the material you are studying is still too difficult for you?

First, try getting a math tutor. Many schools provide tutors through the learning center or math lab. Usually the tutoring is free. If your school does not provide free tutors, then seek out one on your own and pay him or her to help you learn.

There are other things you can do to help yourself if tutoring doesn't work. You can drop the course you are in and sign up for a lower-level course next semester. You can also audit the course, i.e., you can take the course but do not receive a grade. This way you can learn as much as you can but without the pressure. Of course, you will have to enroll in the course again next semester for a grade.

Many colleges and some high schools offer non-credit brush-up courses such as algebra review or arithmetic review. If you need to learn the basic skills, check your school for one of these courses. Usually no tests or grades are given in these courses.

I hope that you have found the suggestions in this program helpful, and I wish you success in your mathematical endeavor.

ABOUT THE AUTHOR

Allan G. Bluman taught mathematics and statistics in high school, college, and graduate school for 39 years. He received his Ed.D. from the University of Pittsburgh and has written three mathematics textbooks published by McGraw-Hill. Mr. Bluman is the recipient of "An Apple For The Teacher Award" for bringing excellence to the learning environment and the "Most Successful Revision of a Textbook" award from McGraw-Hill. His biographical record appears in *Who's Who in American Education, Fifth Edition.*

INDEX

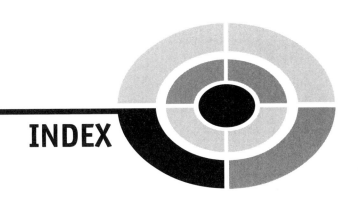